NUREG-1756

I0494047

# Safety Culture: A Survey of the State-of-the-Art

Prepared for the Advisory Committee on Reactor Safeguards

U.S. Nuclear Regulatory Commission
Advisory Committee on Reactor Safeguards
Washington, DC 20555-0001

# AVAILABILITY OF REFERENCE MATERIALS
# IN NRC PUBLICATIONS

# Safety Culture: A Survey of the State-of-the-Art

## Prepared for the Advisory Committee on Reactor Safeguards

Manuscript Completed: December 2001
Date Published: January 2002

Prepared by:
J.N. Sorensen,
Senior Fellow

Advisory Committee on Reactor Safeguards
U.S. Nuclear Regulatory Commission
Washington, DC 20555-0001

# ABSTRACT

This report was prepared for the Advisory Committee on Reactor Safeguards to provide background information on the evolution of the term "safety culture" and the perceived relationship between safety culture and the safety of operations in nuclear power generation and other hazardous technologies. There is a widespread belief that safety culture is an important contributor to the safety of operations. Empirical evidence that safety culture and other management and organizational factors influence operational safety is more readily available for the chemical process industry than for nuclear power plant operations. The commonly accepted attributes of safety culture include good organizational communication, good organizational learning, and senior management commitment to safety. Safety culture may be particularly important in reducing latent errors in complex, well-defended systems. The role of regulatory bodies in fostering strong safety cultures remains unclear, and additional work is required to define the essential attributes of safety culture and to identify reliable performance indicators.

# CONTENTS

## Figures

## Tables

# ABBREVIATIONS

ACSNI       Advisory Committee on the Safety of Nuclear Installations (United Kingdom)
AHP         analytical hierarchy process
ASCOT       Assessment of Safety Culture in Organization Team
ATHEANA     A Technique for Human Event Analysis

BNL         Brookhaven National Laboratory

HSE         Health and Safety Executive (United Kingdom)

IAEA        International Atomic Energy Agency
INEEL       Idaho National Engineering and Environmental Laboratory
INPO        Institute of Nuclear Power Operations
INSAG       International Nuclear Safety Advisory Group

LER         licensee event report

NEA         Nuclear Energy Agency
NOMAC       Nuclear Organization and Management Concept
NRC         U.S. Nuclear Regulatory Commission

OECD        Organization for Economic Cooperation and Development

PNL         Pacific Northwest Laboratory
PRA         probabilistic risk assessment
PRIMA       process risk management audit

SADT        structured analysis and design technique
SALP        Systematic Assessment of Licensee Performance
SKi         Swedish Nuclear Power Inspectorate

TMI         Three Mile Island

WPAM        work process analysis model

# 1. INTRODUCTION

The nuclear industry and the U.S. Nuclear Regulatory Commission (NRC) explicitly recognized the importance of management and organizational factors to nuclear facility safety in the aftermath of the accident at Three Mile Island (TMI) Unit 2. Following the Chernobyl accident, the International Nuclear Safety Advisory Group (INSAG) introduced the term "safety culture" to denote the management and organizational factors that are important to safety [1][1]. Although INSAG intends "safety culture" to capture all the management and organizational factors that are relevant to safe plant operation, many investigators use the term more narrowly. "Safety culture" is often used to denote an element of organizational culture that, in turn, is a component of the broader term "management and organizational factors."

Although major accidents often involve an unsafe act (or failure to act) by an individual, they may also involve conditions created by an organization that can magnify the consequences. The NRC's investigation of the accident at Three Mile Island reported to the Commissioners and the public that "The one theme that runs through the conclusions we have reached is that the principal deficiencies in commercial reactor safety today are not hardware problems, they are management problems" [2]. Later, the report stated, "The NRC, for its part, has virtually ignored the critical areas of operator training, human factors engineering, utility management, and technical qualifications." That sentence captures the basis for much of the NRC's regulatory agenda in the years following the accident, as well as the industry's agenda to improve plant operations.

The NRC's post-TMI action plan included a large number of issues under the general heading of human factors. The major categories included operator qualifications and training, staffing levels and working conditions, the man-machine interface, emergency operating procedures, human reliability, and organizational and management effectiveness. The post-TMI actions included establishing the Systematic Assessment of Licensee Performance (SALP) program in 1980. Its stated purpose was "to provide a means for integration of staff observations, findings and conclusions on licensee performance and management effectiveness" [3].

Independent of the initiatives undertaken by the NRC, the industry saw a need to improve the quality of nuclear operations. The Nuclear Safety Analysis Center was established within the Electric Power Research Institute, and the Institute of Nuclear Power Operations (INPO) was established by the electric utilities that owned and operated nuclear power plants. INPO's task was to foster excellence in plant operations.

Confidence in facility management and human performance within the international nuclear power community was severely damaged by the Chernobyl accident in 1986. In its report of the Chernobyl post-accident review meeting [1], INSAG concluded:

---

[1]Numbers in square brackets identify references listed in Section 10 of this report.

"There is a need for a 'nuclear safety culture' in all operating nuclear power plants.

"The root cause of the Chernobyl accident ... is to be found in the so-called human element. The lessons learned from this imply three lines of action:

(1)     Training, with special emphasis on the need to acquire a good understanding of the reactor and its operation, and with the use of simulators giving a realistic representation of severe accident sequences;

(2)     Auditing, both internal and external to the utility, in particular to prevent complacency arising from routine operation;

(3)     A permanent awareness by all personnel of the potential safety implications of any deviation from the procedures.

"The vital conclusion drawn is the importance of placing complete authority and responsibility for the safety of the plant on a senior member of the operational staff of the plant. Formal procedures properly reviewed and approved must be supplemented by the creation and maintenance of a 'nuclear safety culture.' This is a reinforcement process which should be used in conjunction with the necessary disciplinary measures."

Although the term "safety culture" was introduced after the Chernobyl accident, the underlying concept that an organization's beliefs and attitudes affect its safety performance is much older. Ostram, et al. [4], note that "Heinrich's Domino Theory developed in the 1930s was based on the premise that a social environment conducive to accidents was the first of five dominos to fall in an accident sequence."

The present report explores the nature of safety culture and its perceived importance in the management and regulation of hazardous technologies. The purpose of the report is to provide a tutorial, for non-practitioners of the human performance disciplines, that addresses the following questions:

•       What is safety culture?

•       How can it be measured?

•       How is safety culture related to safety of operations?

•       How is safety culture related to the regulatory process?

Note that addressing these questions does not imply a promise to provide answers to all of them. As one investigator observed, "...the sheer multiplicity of constituent elements of a safety culture and its precept of universal involvement imply that any attempt to monitor its health ... is bound to be complex ..." [5].

Because the term "safety culture" was introduced by INSAG, we first look at INSAG's development of the idea, and the structure it designed for evaluation and implementation. Next, we consider the intellectual foundation of the concept, independent of the INSAG construct. We then discuss the larger issue of human performance, and the place of safety culture within that context. Since the ultimate objective is to establish a relationship between safety culture and the safety of facility operations, we next define the steps required to demonstrate such a link, and review some of the work that has been published toward that end. Finally, we look at the relationship between safety culture and the regulatory process, and identify areas where additional work would appear to be beneficial.

## 2. EVOLUTION OF THE TERM "SAFETY CULTURE"

Having introduced the term "safety culture" into the nuclear safety discussion, INSAG expanded on its importance in INSAG-3, "Basic Safety Principles for Nuclear Power Plants" [6], published in 1988. In that document, INSAG divides safety principles into two categories, fundamental principles and specific principles. The first category, fundamental principles, comprises management responsibilities, the defense-in-depth strategy and general technical principles. The second category, specific principles, includes requirements on siting, design, manufacturing, construction, operation, and accident management

Under the fundamental principle of management responsibilities, INSAG listed three elements: safety culture, responsibility of the operating organization, and regulatory control and independent verification. INSAG also stated a principle for each of these three elements. Among these, the principle stated for safety culture is that "An established safety culture governs the actions and interactions of all individuals and organizations engaged in activities related to nuclear safety." The ensuing discussion explains that:

> "The phrase 'safety culture' refers to a very general matter, the personal dedication and accountability of all individuals engaged in any activity which has a bearing on the safety of nuclear power plants. The starting point for the necessary full attention to safety matters is with the senior management of all organizations concerned. Policies are established and implemented which ensure correct practices, with the recognition that their importance is not just in the practices themselves but also in the environment of safety consciousness which they create. Clear lines of responsibility and communication are established; sound procedures are developed; strict adherence to these procedures is demanded; internal reviews are performed of safety related activities; above all, staff training and education emphasize the reasons behind the safety practices

3

established, together with the consequences for safety of shortfalls in personal performance.

"These matters are especially important for operating organizations and the staff directly engaged in plant operation. For the latter, at all levels, training emphasizes the significance of their individual tasks from the standpoint of basic understanding and knowledge of the plant and the equipment at their command, with special emphasis on the reasons underlying safety limits and the safety consequences of violations. Open attitudes are required in such staff to ensure that information relevant to plant safety is freely communicated; when errors of practice are committed, their admission is particularly encouraged. By these means, an all pervading safety thinking is achieved, allowing an inherently questioning attitude, the prevention of complacency, a commitment to excellence, and the fostering of both personal accountability and corporate self-regulation in safety matters."

At this point in its evolution, the term "safety culture," as described in INSAG-3, is not clearly distinguished from all of the other functions or attributes that contribute to nuclear safety. Once it has been said that the "starting point for the necessary full attention to safety matters is with the senior management of all organizations concerned," the requirements that follow are part of the collection of preferred practices identified before the term "safety culture" was introduced. If anything is added, it is the idea that "... their importance lies not just in the practices themselves, but also in the environment of safety consciousness which they create." This assertion seems plausible, but it probably requires testing in some way.

The discussion of defense in depth as a fundamental safety principle in INSAG-3 mentions "safety culture" as one of the human aspects of defense in depth, along with quality assurance, safety reviews, and personnel qualifications and training. Safety culture is again mentioned in connection with the accident prevention aspect of defense in depth. Specifically, "In accordance with the general safety management principle on safety culture, the safety implications of decisions [regarding design, construction, operation and maintenance] must be borne in mind."

INSAG-3 mentions safety culture in three other places. The first is in the discussion of design management, where it states that "The design of a safe plant is under the authority of a highly qualified engineering manager whose attitudes and actions reflect a safety culture and who ensures that all safety and regulatory requirements are met." In the discussion of operational limits and conditions, INSAG-3 states that "As a vital part of safety culture, it is essential that plant personnel understand the reason for the safe limits of operation and the consequences of violation. The third instance is in an appendix discussing defense in depth, which states that "A first level of protection in defence in depth is a combination of conservative design, quality assurance, surveillance activities, and a general safety culture that strengthens each of the successive obstacles to the release of radioactive materials."

In INSAG-3, it seems that INSAG is still searching for a clear definition and a clear role for safety culture in nuclear safety. Safety culture is not defined in either INSAG-1 or INSAG-3, except in an operational sense. Its precise contribution to safety is unknown, and it could easily be dismissed as another one of those things that seem like a good idea.

## 3. SAFETY CULTURE DEFINED

In 1991, INSAG published INSAG-4, "Safety Culture" [7], which deals exclusively with safety culture, how it is defined, and how it might be assessed. The Foreword to INSAG-4 notes the introduction of the term in INSAG-1, its expansion in INSAG-3, and its subsequent increased use in the nuclear safety literature. It then states, "However, the meaning of the term was left open to interpretation and guidance was lacking on how Safety Culture could be assessed."

INSAG-4 defines safety culture as "... that assembly of characteristics and attitudes in organizations and individuals which establishes that, as an overriding priority, nuclear plant safety issues receive the attention warranted by their significance." It then explains that "This statement was carefully composed to emphasize that Safety Culture is attitudinal as well as structural, relates both to organizations and individuals, and concerns the requirements to match all safety issues with appropriate perceptions and action."

Since the definition of safety culture is related to personal attitudes and habits of thought, as well as to the style of organizations, INSAG-4 suggests that "... such matters are generally intangible; that nevertheless such qualities lead to tangible manifestations; and that a principal requirement is the development of means to use the tangible manifestations to test what is underlying." Arguing that "... sound procedures and good practices are not fully adequate if merely practised mechanically... ," INSAG-4 holds that "... Safety Culture requires all duties important to safety to be carried out correctly, with alertness, due thought and full knowledge, sound judgement and a proper sense of accountability."

The body of INSAG-4 is devoted to articulating what INSAG terms "universal features of a safety culture" and identifying broad characteristics (tangible evidence) of an effective safety culture. The approach to both topics, universal features and tangible evidence, is to provide detailed lists of the desired attributes. This approach is reminiscent of INSAG's approach to defense in depth, where it provides a complete structure for defense-in-depth features ranging from design through operation and from component functionality through emergency planning.

The universal features of safety culture are divided into three broad categories, which encompass requirements at the policy level, requirements on managers, and the response of individuals. The top-level requirement is that "Governments discharge their responsibilities to regulate the safety of nuclear plants ... in order to protect individuals, the public at large, and the environment. Legislation is backed by the necessary advisory and regulatory bodies, which have sufficient staff, funding and powers to perform their duties and the freedom to do so without undue interference." Additional policy-level requirements include statements of safety policy, management structures that provide accountability in safety matters, adequate resources devoted to safety, periodic self-assessment by all organizations with safety responsibility, and a visible commitment to safety by senior managers. The role of the regulatory body is addressed with a requirement that "... an effective Safety Culture pervades its own organization and its staff. The basis is ... a safety policy statement."

Requirements imposed on managers include providing clear lines of responsibility and authority, defining and controlling work practices, ensuring appropriate qualifications and adequate training for staff, and providing a system of rewards and sanctions that promotes good safety practices.

The responsibility imposed on individuals is summarized as maintaining a questioning attitude, adopting a rigorous and prudent approach to safety-related tasks, and maintaining communications that contribute to safety. The questioning attitude is characterized by questions, such as "Do I understand the task?" and "Are there any unusual circumstances?" The rigorous and prudent approach includes understanding procedures, complying with procedures, and seeking help if necessary. Communications includes obtaining information from others, transmitting information to others, and reporting on and documenting results of work.

Although INSAG-4 devotes particular attention to operating organizations, the discussion is intended to extend the concept of safety culture to any organization that can affect safety, including the functions of design and safety research. Safety culture is considered as "... the assembly of commendable attributes of any organization or individual contributing to nuclear plant safety." While acknowledging that the attributes of safety culture are, for the most part, intangible, INSAG deems it important to be able to judge the effectiveness of safety culture. It develops the framework for judging effectiveness by asserting that intangible attributes produce tangible manifestations that can act as indicators. It then identifies the manifestations expected, or hoped for, in government, operating organizations, and supporting organizations.

Government commitment should be evident in legislation and policies that set broad safety objectives, establish necessary institutions, and ensure adequate support. Commitment should also be evident in the relationship between the regulatory organization and the operating organization. Relationships should be open, but with sufficient formality to ensure accountability. Regulators should recognize that primary responsibility for safety lies with the operating organization. Adopted standards should establish appropriate levels of safety, while recognizing the inevitable residual risk.

Within the operating organization, INSAG looks first at the corporate policy level, stating that "Safety Culture flows down from actions by the senior management of an organization... The primary indication of corporate level commitment to Safety Culture is its statement of safety policy and objectives." Other indicators of safety culture should be found in regular reviews of the organization's safety performance and the evaluation of individual attitudes toward safety as part of the staff selection and promotion process.

To find tangible evidence of safety culture among the operating personnel of a particular power plant, INSAG suggests that the three aspects to be considered are (1) the environment created by local management, (2) the attitudes of individuals at all levels, and (3) the actual safety experience at the plant. The working environment should include defined safety responsibilities and detailed practices at all levels. Training and education should ensure staff knowledge about possible errors in each individual's area of activity. Safety concerns should be given a high level of visibility by plant inspections, audits, visits by senior officers, and safety seminars. Satisfactory facilities, including tools, equipment and information, should be provided to the staff.

Individual attitudes are reflected by adherence to procedures, stopping to think when facing an unforseen situation, and management respect for a good safety attitude. Managers should take opportunities to show that they will put safety concerns ahead of power production if circumstances warrant. Development of local practices for enhancing safety, such as error reporting, should be encouraged.

Ultimately, in INSAG's view, the effectiveness of the organization's safety culture should be reflected in the performance of the facility. Plant performance indicators, including plant availability, number of unplanned shutdowns, or radiation exposure, are a reflection of attention to safety. Significant events that occur should be analyzed to determine what they reveal about staff strengths and weaknesses. The rigor of the reviews, and the effectiveness of any resulting corrective actions, are important safety culture indicators.

The final area treated in INSAG-4 is supporting organizations, specifically research or design organizations. The treatment is cursory, and does little more than endorse the importance of safety culture at such institutions. Research organizations should monitor relevant work around the world that might presage new safety issues. Design organizations should keep up to date on reactor safety technology developments.

The conclusion presented in INSAG-4 is that "safety culture" is now a commonly used term, and that it is important to give practical value to the concept. This includes identifying attributes that may be used to judge the strength of safety culture in specific instances. In addition, INSAG-4 includes as appendix that identifies questions that INSAG suggests are worth examining when assessing the effectiveness of safety culture in a particular situation. The questions are organized along the same outline as the body of the report, covering government, operating, research, and design organizations. By far, the most extensive coverage is given to operating organizations, which are the focus of 73% of the 143 questions. Less coverage is given to government organizations, which are the focus of 22% of the questions. As with the body of the report, the attention to research and design organizations, each with 6% of the questions, is somewhat cursory.

Under the heading of government commitment to safety, typical questions include: Is the body of legislation satisfactory? Is funding sufficient to allow hiring staff of adequate competence? Are there any instances of undue interference in technical matters with safety relevance? Regarding the performance of regulatory agencies, questions include: Are regulatory safety objectives annunciated clearly? What is the record of project delays or loss of production due to lack of clarity of regulatory requirements or lack of timely regulatory decisions? Is there mutual respect between the regulatory staff and the operating organization based on a common level of competence?

As noted above, the majority of the suggested questions are directed at the operating organization, including: Has a safety policy statement been issued? Is there an active nuclear safety review committee that reports its findings at the corporate level? Do the staff recognize that attitude toward safety is important in the selection and promotion of managers?

Following the publication of INSAG-4, the International Atomic Energy Agency (IAEA) published guidelines [8] "... for use by any organization wishing to conduct a self-assessment of safety culture." Entitled "ASCOT Guidelines," (Assessment of Safety Culture in Organizations Team Guidelines), the document summarizes the concept of safety culture and then describes a process for assessing safety culture. It includes how the IAEA can support such an assessment, ranging from providing advisory services to actually performing the assessment.

Safety culture is considered to have two major components, which are (1) the environment created within which personnel work, and (2) the attitude and response of individuals to that environment. The ASCOT guidelines restate the basic INSAG questions and expand on them with approximately 300 "Guide Questions." For example, the basic question from INSAG-4, "Are there any undue impediments to the necessary amendment of regulations?", is supplemented with the guide question, "What is the mechanism and how long does it take to make changes to your nuclear legislation?". In addition, each group of basic questions and guide questions is keyed to the organizational levels that should be asked to respond to those questions. To continue the present example, the ASCOT guidelines suggest addressing questions on the adequacy of legislation to utility corporate management, operating organization (plant) management, and regulatory or government personnel. Other groups of questions are addressed to individuals at the plant and to support organizations.

Following each group of guide questions is a list of key indicators that evaluators should look for. In the case of satisfactory government support, plant staff and regulators should confirm that there is no political interference in safety matters, and that the regulator has adequate manpower and enforcement rights.

The ASCOT guidelines suggest that a practical assessment of safety culture include a plant walk-through and an overview of plant documentation. The walk-through should include observations of access control, general plant condition, housekeeping, use of protective equipment, alertness of control room staff, and availability of procedures and manuals. The documentation review should include log-books, operations and maintenance records, training program descriptions, safety policies, plant policies on adherence to procedures, organization charts and job descriptions, and documents identifying key safety responsibilities.

The body of the safety culture review consists of addressing the basic questions and guide questions to the appropriate groups within the scope of the evaluation. Typically, the scope would include regulators, corporate management, plant management, individual plant workers, and supporting organizations. The ASCOT guidelines emphasize the need to look for "... tangible evidence of an essentially intangible concept." They illustrate this by suggesting that an evaluation of an audit program should go beyond the review of audit reports and corrective actions. Safety culture can be better evaluated by looking at the underlying attitudes. For example, do managers show support for audits to their staff? Are auditors considered to be technically competent? Are corrective actions taken enthusiastically?

Missing from the ASCOT guidelines, as well as from INSAG-4, is any indication of how an overall conclusion should be drawn from the collected answers to all the questions. Possibly, the intent of judging safety culture does not include an overall conclusion. It may be that the intent is simply to identify deficiencies and make suggestions for improvements in each area. Still, it would seem that a facility with a poor safety culture might be left with an overwhelming list of corrective actions. Unless some guidance is provided on how to proceed, the evaluation may provide little help. At the other extreme, a facility with an exceptionally good safety culture should be so recognized and given positive reinforcement. It would seem inevitable that the review team will conclude that a safety culture is superior, acceptable, or deficient, and attempt to provide the proper degree of motivation for corrective actions.

The fundamental problem with INSAG's approach to safety culture is that it specifies in great detail what should be included, but provides little guidance on overall criteria for acceptability. Furthermore, no link is made (or even seems possible) between safety culture as INSAG defines it and human performance or human reliability. A positive relationship is simply assumed. One of the goals to be reached in risk-informed regulation is to advance the state-of-the-art in probabilistic risk assessment (PRA) to account for the probability of human error and the contribution of human skills to recovering from accident sequences. The INSAG approach appears to make little contribution to either aspect. Similarly the INSAG work does not establish the link between a good safety culture and safe plant performance. Again, the relationship is simply assumed. While it seems plausible that the sum total of the indicators of a strong safety culture would imply safe plant operations, that is not the same as demonstrating a "cause and effect" relationship. The possibility remains that safe plant operations can be fostered, perhaps even more effectively, by other organizational characteristics.

## 4. ORGANIZATIONAL CULTURE

Although INSAG has borrowed the term "culture" from either anthropologists or the organizational development community (who, in turn, borrowed it from anthropologists), the INSAG publications make no reference to the bodies of literature in those fields. In fact, no attempt is made to link "safety culture" with "culture" as the term is used elsewhere. In *An Introduction to Cultural Theory and Popular Culture* [9], John Storey avers that "In order to define popular culture we first need to define the term culture." Similarly, it would seem that in order to define "safety culture" it would be necessary to first define "culture."

Storey goes on to suggest three broad definitions. The first is that culture refers to "a general process of intellectual, spiritual, and aesthetic development." The second is that culture might identify "a particular way of life, whether of a people, a period, or a group." A third definition could be "the works and practices of intellectual and especially artistic activity." Of these three suggestions, the second seems to best support the extension of "culture" to "safety culture."

It is also possible that INSAG looked to the organizational development discipline as supporting the choice of the word "culture" to describe the organizational attributes it found desirable relative to safety. The idea that the term "culture" could be used to describe some attributes of an organization began to appear in the early 1980s. The implication of adopting this term was that by associating the attributes of interest with an organizational culture, one could gain insights about the behavior of the organization or its members. These insights, in turn, could be used to guide the management of the organization in establishing and reaching organizational goals.

An early (1982) book on this subject was *Corporate Cultures* by Terrence Deal and Allen Kennedy [10]. One of the driving forces behind this book was trying to establish why the structure of an organization often did not explain its control of work activities. In developing the idea of corporate culture, the authors started by defining culture according to Webster's New Collegiate Dictionary, as "the integrated pattern of human knowledge, belief, and behavior that depends upon man's capacity for learning and transmitting knowledge to succeeding generations; the customary beliefs, social forms, and material traits of a racial religious, or social group." They also attributed to Marvin Bower, a managing director of McKinsey & Company, a less formal definition of the cultural elements of business, namely "the way we do things around here." This wording reappears in more recent writings on safety culture.

In 1983, after several books had been published on the human underpinnings of business, Bro Uttal, writing in *Fortune* magazine [11], attempted to summarize the meaning of organizational culture. He defined organizational culture as a system of shared values (what is important) and beliefs (how things work) that interact with a company's people, organizational structures, and control systems to produce behavioral norms (the way we do things around here). This definition was later used by Reason [12], and variations of it appear in a number of papers.

It is important to recognize that culture, including organizational culture, may have some characteristics that are not desirable. For example, culture is difficult to change. That difficulty at least highlights the challenge that organizations face when their safety cultures needs improvement. Some writers argue that, "Culture may simply exist" [11, p. 72].

Jacobs, et al. [13], defined organizational culture as the plant personnel's shared perception of the organization, including traditions, values, customs, practices, goals, and socialization processes that endure over time. "It defines the "personality" of the organization." At least one writer in the general field of organizational behavior, William Bridges [14], would disagree, and argues that "culture" and "personality" are distinctly different. In *The Character of Organizations*, Bridges raises a cautionary note regarding the current practice of assuming an organizational culture exists, can be reasonably well defined, and can be changed. He observes that there are several important differences between "culture" as commonly used by anthropologists and "culture" as applied to organizations by management consultants. He notes that, "Like many who borrow concepts from other fields, organizational writers have oversimplified matters to such an extent that their concept has lost much of its connection to the usages that are current in the field to which it belongs."

Since INSAG introduced the term "safety culture," it has been adopted in the literature on the safety of other activities, including aviation, chemical processing and rail transportation. Writing on creating a safety culture in aviation, Merritt and Helmreich [15] began on a somewhat skeptical note, saying "What is a safety culture, and how does it differ from a safety initiative or a safety system? ... [I]t is necessary to define culture and understand its parameters if one is ever to create or sustain a safety culture."

Apostolakis and Wu [16] questioned whether the term "safety culture" is appropriate, suggesting that it is too narrowly drawn. "When the subject is culture, we must question the wisdom of separating safety culture from the culture that exists with respect to normal plant operation and power production. The dependencies between them are much stronger because they are due to common work processes and organizational factors."

If the culture of an organization is the system of shared values and beliefs that interact with the organization's people, structures, and control systems to produce behavioral norms (the way we do things around here), then safety culture can perhaps be described as the shared values and beliefs on risks, accidents, health, and safety. INSAG appears to intend that its concept of safety culture be expanded as circumstances dictate to cover important organizational influences on safety. Indeed, INSAG's list of questions for judging safety culture carries the caveat that "... the list of questions cannot be comprehensive, nor can a list which is at all extensive be applicable to all circumstances. The objective ... is to encourage self-examination in organizations and individuals rather than to provide a checklist ..."[7].

Despite the reservations of some investigators, safety culture seems to be accepted as an appropriate and useful concept, even though its relationship to culture in the usual sense is tenuous. Ascribing the usually understood characteristics of culture to safety culture should be done with some caution. It should also be noted that the definition of safety culture by INSAG and others is probably incomplete. The term itself implies that it is a subset of a larger organizational culture. Indeed, some writers simply accept safety culture as a component of organizational culture without trying to define either term [17].

Safety culture may not capture all of the management and organizational factors that are important to safe plant operation, but it has acquired a place in the literature. Although the literature does not support any single definition of safety culture, it is probably reasonable to settle on a model that represents organizational culture as a particular application of the larger concept of culture, and then considers safety culture as a subset of organizational culture. The definition chosen for "safety culture" should be consistent with its parent terms, "culture" and "organizational culture." The ultimate objective is to establish a link between safety culture and safety of operations. That process requires not only a definition, but also a delineation of the characteristics or attributes of safety culture. Such attributes should be consistent with the chosen definition, but they are probably more important than the definition. Possible attributes, and their importance in linking safety culture to safety of operations, are discussed later.

## 5. SAFETY CULTURE IN CONTEXT

Safety culture, however defined, is part of the larger issue of human factors. In a 1988 study requested by the Nuclear Regulatory Commission, the National Research Council recommended a human factors research agenda to be undertaken by the Nuclear Regulatory Commission [18]. The recommended program included five major areas: (1) human-system interface design, (2) the personnel subsystem, (3) human performance, (4) management and organization, and (5) the regulatory environment. The first two areas are primarily related to system design and personnel training, respectively, and are only indirectly related to safety culture. The next two areas, human performance and management and organization, are most closely related to the idea of safety culture. Under human performance, the National Research Council identified the highest priority topic as causal models of human error. Under management and organization, it identified two high priority topics, specifically the impact of regulations on the practice of management, and organizational design and a culture of reliability. Equating "culture of reliability" to what we now call "safety culture" seems like a reasonable step.

Safety culture is also related to the last area mentioned by the National Research Council, regulatory environment, but not in a straightforward way. Regulatory activities influence the overall environment in which licensee organizations operate and hence affect the organizational cultures that evolve. Regulatory activities also have the potential to be counterproductive, especially if they appear to shift the responsibility for safety from the operator to the regulator. The INSAG model of safety culture emphasizes both the context created by the organization, and the response of individuals within that context.

### 5.1 Human Error

The focal point of human factors concerns is the performance of individuals. The other four factors identified by the National Research Council provide the context in which the individual functions. The term "human error" is generally understood to mean an unsafe act by a system operator. The consequences of such an act may or may not be severe, depending on other circumstances. Such "other circumstances" are often the product of organizational factors that establish other important conditions that determine system response. Reason [19] distinguishes between active errors, "whose effects are felt almost immediately," and latent errors, "whose adverse consequences may lie dormant within the system for a long time ..." Active errors are usually associated with system operators such as airplane pilots, air traffic controllers, or power plant control room personnel. Latent errors are normally associated with personnel who are removed from operations, such as designers and maintenance personnel.

Reason's human error model addresses the area that the National Research Council termed "human performance." He divides "unsafe acts" into two classes: "unintended actions" and "intended actions" (see Figure 1). Intended actions, in turn, are divided into "mistakes" and "violations." A mistake is an unintentional deviation from a prescribed course of action, while a violation is an intentional deviation. Unintended actions are classed as "slips," "lapses" or "mistakes." A slip involves attentional failures, such as misordering or mistiming intended actions. Lapses are memory failures, such as omitting planned items or forgetting intended actions. Mistakes are classed as either rule-based (misapplication of a good rule or application of the wrong rule), or knowledge-based (inaccurate mental model of the problem).

Modeling human error, as outlined above, is necessary to the complete understanding of the human contribution to system safety. Information from human error models and associated data gathering are an important input to the process of probabilistic risk assessment. The probability of an operator committing an error and causing a system to fail to perform its intended function is as important as a component failure leading to the same result. Modeling unsafe acts, however, is only part of the story. The consequences of those acts often depend on latent errors. It seems reasonable to expect that safety culture, and probably other organizational factors, will have a significant influence on both the frequency of unsafe acts and the probability of latent errors.

According to Reason [19], "There is a growing awareness within the human reliability community that attempts to discover and neutralise these latent failures will have a greater beneficial effect upon system safety than will localised efforts to minimise active errors." The idea that correcting latent errors will have a greater effect on safety of operations than preventing active errors is engaging, but the relative contribution of latent and active errors to safety of operations probably needs to be demonstrated with some degree of rigor. At the least, it is reasonable to expect that the relative contributions will depend on the characteristics of the systems or processes being examined.

5.2 Organizational Accidents

In *Human Error* [19], Reason argues that most of the root causes of serious accidents are present within the system long before an obvious accident sequence can be identified. He contends that "...some of these latent failures could have been spotted and corrected by those managing, maintaining and operating the system in question." In a subsequent book, *Managing the Risks of Organizational Accidents* [12], he looks at the organizational functions involved in creating or

# Figure 1
## Classification of Unsafe Acts
### (from Reason, *Human Error* [19])

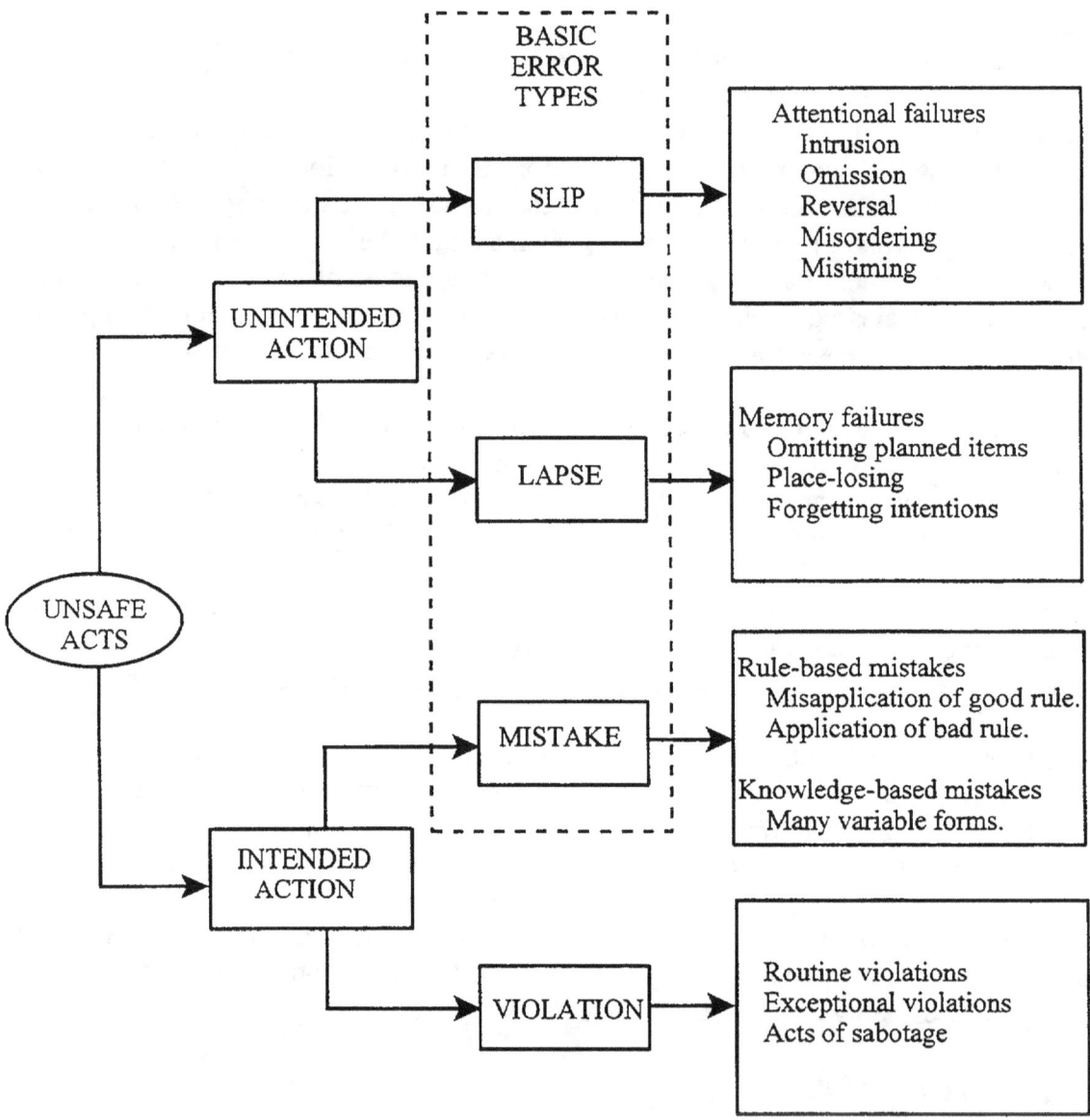

A summary of the psychological varieties of unsafe acts, classified initially according to whether the act was intended or unintended and then distinguishing errors from violations.

14

mitigating accidents. He argues that "...human error is a consequence, not a cause. Errors ... are shaped and provoked by upstream workplace and organizational factors" [12, p.126]. It follows that understanding the management and organizational factors that can either reduce or identify and correct latent errors is an important element in reducing the frequency and consequences of accidents.

Typically, organizational accidents involve "...the interaction of latent conditions with local triggering events" [12, p.35]. Reason describes organizational accidents in terms of organizational factors, local workplace factors and unsafe acts. The organizational factors and local workplace factors not only interact directly, but each may create latent condition pathways. Accidents with significant losses occur when all of these conditions align in such a way that the defenses built into a system are overwhelmed.

Reason maintains that latent conditions may be sufficient to cause accidents, and that they are always present in the system. He then notes that the quality of both production and protection depend on the same organizational processes. Furthermore "... the partnership between production and protection is rarely equal ... partly because those who manage the organization possess productive rather than protective skills, and partly because the information relating to production is direct, continuous, and readily understood [12, p.4]." By contrast, "... safe operations generate a constant — and hence relatively uninteresting — non-event outcome. The mechanism by which this reliability is achieved can be opaque to those who operate and manage the system" [12, p.37].

In May 2000, the NRC issued a report describing a human reliability analysis method called A Technique for Human Event Analysis (ATHEANA) [20]. The model provides a structured search process for human failure events, including detailed search processes for error-forcing context, and an improved representation of human-system interactions. The issues addressed by the concept of safety culture in general, and latent errors in particular, provide what is called the "error forcing context" for ATHEANA.

The ATHEANA process contributes to the objective of systematically identifying important management and organizational factors that contribute to significant event sequences. The ATHEANA analysis of the Wolf Creek drain-down event [21] identified a number of management and organizational factors that contributed to the occurrence of the event. These contributors included incompatible work activities, a compressed outage schedule, poor mental models of the systems and valves, heavy reliance on the control room crew to identify problems, and inadequate reviews of procedures prior to use.

The influence of latent errors was identified in a recent study by the Idaho National Engineering and Environmental Laboratory (INEEL) [22]. One objective of the study was to identify the influence of human performance in significant operating events. INEEL analyzed 35 operating events, 20 of them using PRA methods. Event importance, as measured by conditional core damage probability, ranged from $1.0 \times 10^{-6}$ to $5.2 \times 10^{-3}$.

INEEL found that most identified errors were latent, with no immediate observable impact. The ratio of latent to active errors was 4:1. Latent errors included failure to correct known problems, incomplete design change testing, inadequate maintenance practices and post-maintenance testing, and poor work package quality assurance. Active errors included failures in command and control (such as loss of phone communications), and incorrect operator actions (such as incorrect line-ups or acting without procedural guidance).

The INEEL findings are supported by other analyses. In discussing a human performance improvement program at Duke Power Company, one Duke Power senior manager observed that "If you analyze an entire event, ... you'll find it wasn't just one mistake — it was five, six or seven mistakes that occurred and there weren't enough contingencies or barriers built in to prevent the event from happening" [23].

A systematic effort to improve human performance at Duke Power's McGuire nuclear power station, which addresses virtually the same factors identified by INSAG's model of safety culture, has produced significant improvements in station performance [23]. The McGuire program was started in 1994 when declining performance required correction and management determined that station processes and programs were to blame. Similar programs were later started at other Duke Power stations, and were brought under corporate management direction by 1996.

A structured assessment by Duke of human performance needs identified the need for focused human error reduction training for technicians and supervisors. Although the term "safety culture" is not used in describing the program, it incorporates elements and issues that are practically identical to many of those addressed by INSAG-4. One element in the Duke Power program, for example, is "individual commitment," which includes a questioning attitude, procedure use and adherence, communications, stopping when unsure, and an overall prudent approach. The same parallels exist for the manager's commitment portion of the INSAG model and the supervisor's and manager's sections of the Duke Power program. Both deal with clear priorities, goals, and responsibilities, clear lines of responsibility and authority, staff skills and competence, and performance assessment.

Since the program was initiated, refueling outage times at McGuire have been reduced from about 90 days to about 33 days, and capacity factors have increased from about 72% to about 89%. These results, of course, are measures of efficiency, not safety. Nonetheless, the similarity between the management and organizational factors apparently responsible for the noted improvements and those factors identified with safety culture suggests that an attempt to relate "safe operations" to "efficient operations" might be worthwhile. It is often claimed that facilities that are efficient and well managed from a production standpoint are also safe facilities. That notion is not universally accepted, and probably requires a more rigorous examination than it has received to date. Such an examination may be valuable.

## 6. RELATING SAFETY CULTURE TO SAFETY PERFORMANCE

As noted previously, one of the omissions in INSAG's structure for establishing and evaluating safety culture is the link between safety culture and safety of operations. The INSAG approach assumes, but does not attempt to demonstrate, a positive relationship between safety culture and facility safety. There are actually two parts to this demonstration. The first part is to establish a relationship between safety culture (or its associated attributes) and safety of operations. The second part is to determine whether there are suitable performance indicators that can be used to infer changes in safety culture and, thereby, predict changes in safety performance. There is a substantial body of literature that addresses the first part of the problem. There is much less work that addresses the second part, however. No performance indicators to gauge safety culture and its impact on safety of operations appear to have been identified and validated.

Before reviewing the literature, it may be useful to discuss the logical framework required to demonstrate that an effective safety culture results in improved operational safety. (This framework facilitates discussion of the relevant research in the next section of this report.)

Figure 2 shows an activity diagram for establishing a relationship between safety culture and safety, the first part of the problem posed above. The objective is to identify one or more measurable attributes of safety culture that can be correlated with one or more measures of operational safety. The second part of the problem, identifying suitable performance indicators, is outlined in Figure 3.

Research intended to show how management and organizational factors affect safety of operations typically begins by describing how a particular organization works, and attempts to identify specific, measurable organizational factors that influence safety. The process necessarily requires some measure of safety, such as the frequency of accidents. The analyst may begin by choosing an organizational model to represent how the organization works. The insights derived from that model, in conjunction with a suitable definition of safety culture, can be used to suggest attributes of safety culture that can be measured (step 1 in Figure 2). Such attributes might include, for example, effectiveness of organizational communications, organizational learning, management attention to safety, and management expectations regarding compliance with procedures.

The next step in the process is to design methods to measure the proposed attributes in a real organization. This typically involves using audits, inspections, document reviews and personnel surveys. The tools and techniques used here often include those used by psychologists as well as those used by engineers. To continue the example, designing the measurement methods involves finding a way to quantify "management attention to safety" and the other proposed attributes of

Figure 2
Relating Safety Culture to
Safety Performance

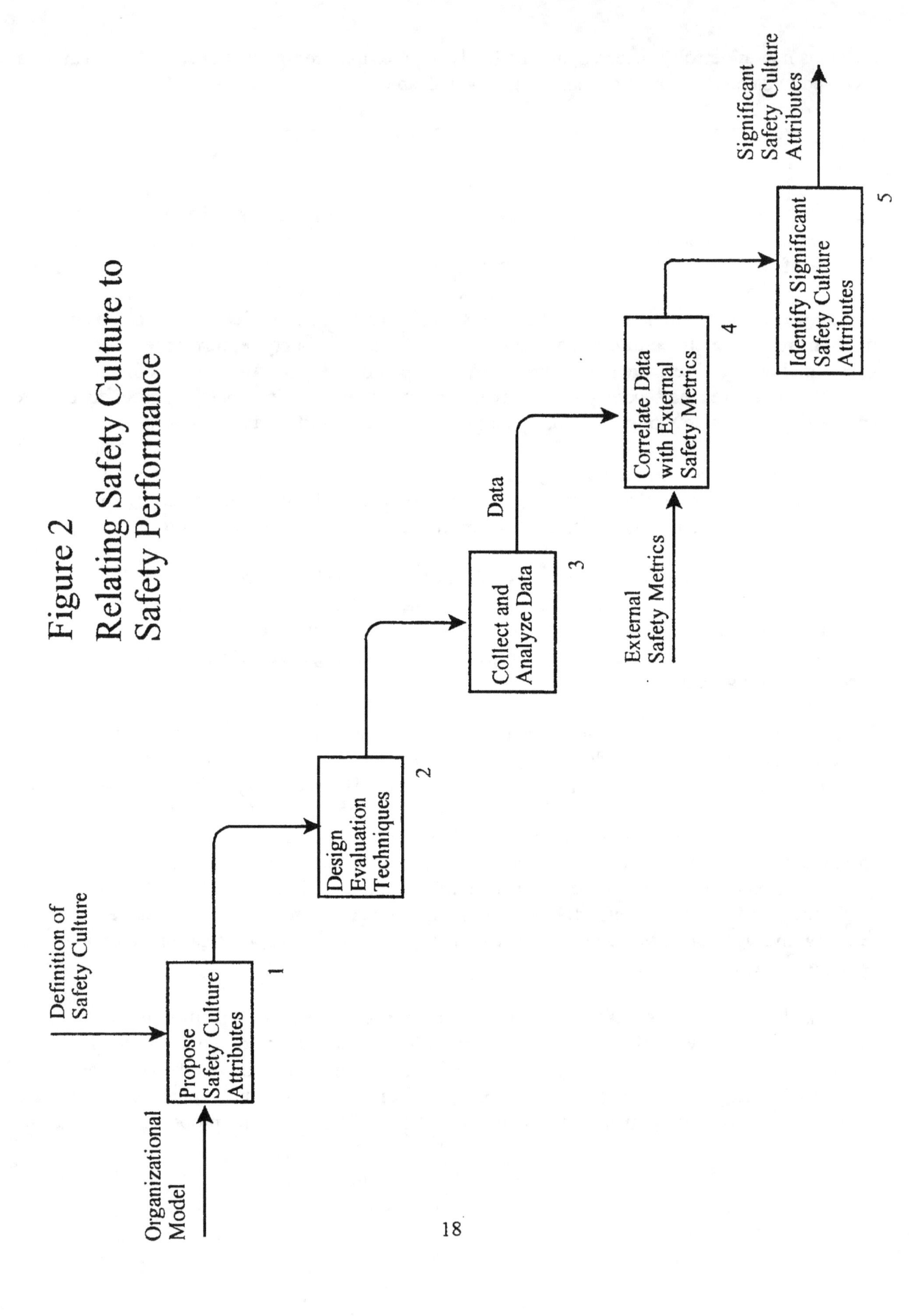

Organizational
Model

Definition of
Safety Culture

Propose
Safety Culture
Attributes

1

Design
Evaluation
Techniques

2

Collect and
Analyze Data

3

Data

External
Safety Metrics

Correlate Data
with External
Safety Metrics

4

Identify Significant
Safety Culture
Attributes

5

Significant
Safety Culture
Attributes

18

Figure 3
Relating Safety Culture to
Risk Metrics

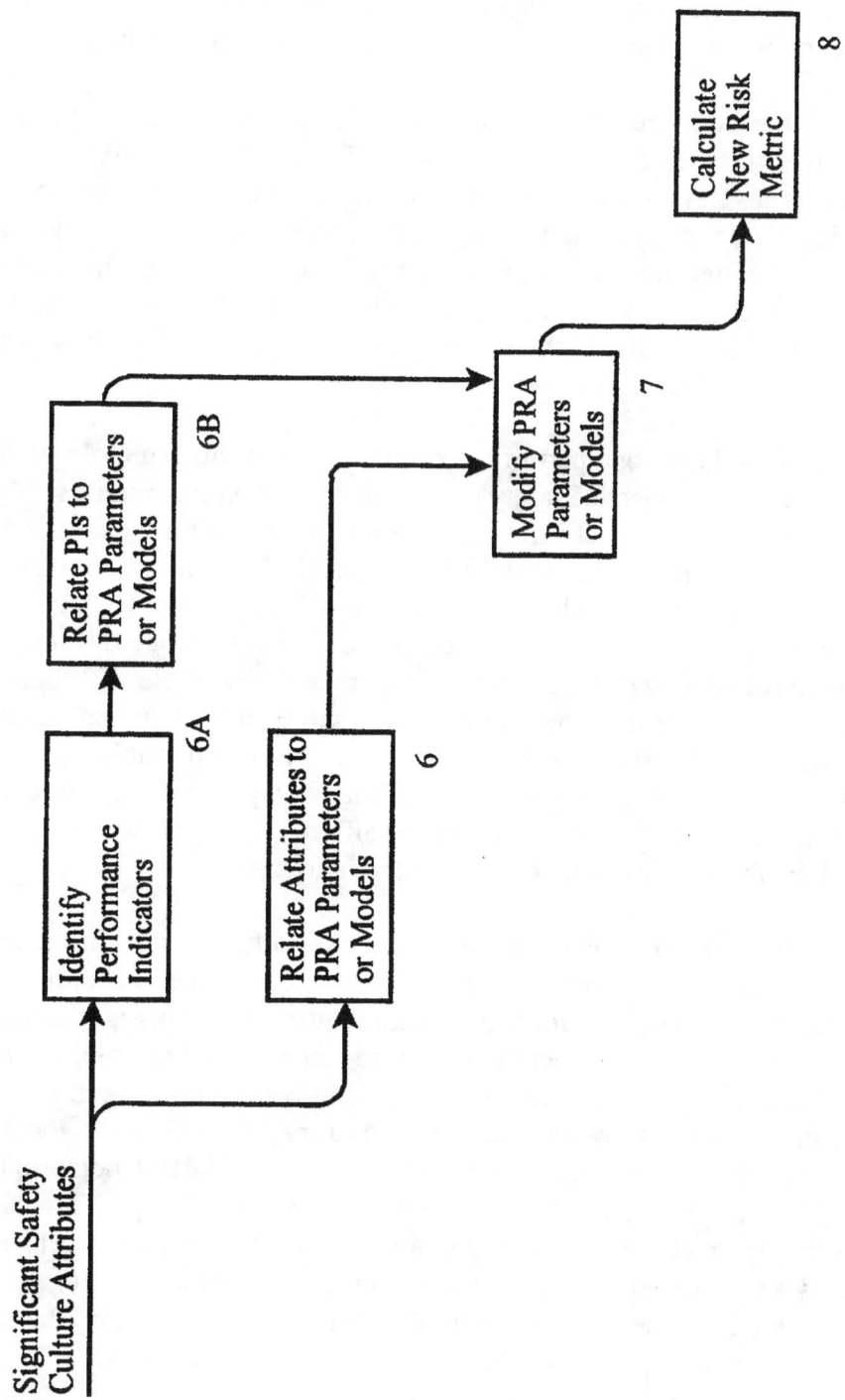

safety culture. Selection of the measurement techniques is obviously followed by data collection and analysis (step 3 in Figure 2).

The next step is to correlate the attribute measurements with one or more measures of operational safety. This step obviously requires the analyst to select external safety metrics, and the choices may be dictated by what measures are available. Early studies of U.S. nuclear power plants used SALP evaluations, licensee event reports (LERs), and other performance indicators such as unplanned scrams, safety system actuations, and safety system failures.

Correlation of the safety culture attribute data with the chosen safety metrics (step 4 in Figure 2) usually involves the use of regression analysis. The results typically show that some of the proposed attributes have a statistically significant relationship with one or more of the chosen safety metrics, while other attributes may show no significant correlation. An organization with a low score on "management attention to safety," for example, might consistently have a relatively high rate of safety system failures. The output of the process at this point (step 5 in Figure 2) is the identification of those safety culture attributes that show a significant relationship to safety, at least as measured by the chosen safety metrics.

Figure 3 addresses the second part of the problem, identifying suitable performance indicators for safety culture and the impact of safety culture attributes on risk metrics. As shown by the lower path in Figure 3, the significant safety culture attributes must be related to parameters in a PRA, such as human error probability, system failure probability, or system unavailability. Essentially, a numerical value must be developed for each significant element. An algorithm is then developed to relate the resulting quantification to a change in one or more PRA parameters, such as a system unavailability or failure rate. To pursue the example of "management attention to safety," the desired algorithm could relate a low score on this attribute to an increase in assumed equipment failure rates used as input to the PRA. It is also possible that a relationship identified between the significant attributes and the external safety metrics is not modeled in the PRA at all. In this case, the PRA model itself must be modified. The final step in this path is the calculation of core damage frequency or other chosen risk metric.

For the overall process described above to be most useful in assessing the safety of hazardous facilities, it is desirable to identify easily obtainable performance indicators that will provide a reliable measure of the significant safety culture attributes. This is illustrated in the upper path in Figure 3. Evaluating "management attention to safety," for example, might require extensive data collection and analysis. Once the relationship between management attention to safety and safety system failure rates has been established, it may be possible to identify an easily observable surrogate for management attention to safety. Such a performance indicator might be the fraction of employees participating in periodic safety training. This indicator could be monitored through record reviews, and would not require the personnel surveys and audits that might otherwise be needed to measure management attention to safety. If suitable performance indicators for the attributes of safety culture can in fact be identified, the performance indicators can also be related, in turn, to the PRA parameters or PRA models.

# 7. MODELING ORGANIZATIONS

This section reviews some of the literature that addresses studies related to establishing a relationship between safety culture (or other management and organizational factors) and safety of operations. The models discussed in each of the studies reviewed address some of the activities represented in Figures 2 and 3. None of the models treats all activities, at least not with the same degree of thoroughness and rigor. Typically, a given study addresses a few activities in detail, and acknowledges the need to address the remainder. Some studies consider safety culture to be a subset of the management and organizational factors that might affect safety, and others do not use the term "safety culture" at all. The studies selected for review are a representative, but limited, sample of the available literature, not an exhaustive survey.

## 7.1 Chemical Industry Safety Surveys and Audits

Investigators in the chemical process industry have used safety audits and personnel surveys as the primary means of relating safety attitude or safety culture to operational safety. Investigators in this field have the advantage (if it can be deemed an advantage) that certain types of accidents occur with sufficient frequency to provide statistically valid measures of operational safety.

Donald & Canter [24] examined the relationship between employee attitudes and safety performance in the chemical process industry using the terms "safety attitudes" and "safety climate" instead of "safety culture." The authors use the term "organizational climate" as the sum of the perceptions that employees have of their organizations. The climate represents the context in which behavior occurs and the basis of people's expectations.

Donald and Canter began their study by deriving from the relevant literature six factors that are associated with safety:

- management commitment
- safety training
- open communication
- environmental control and management
- stable workforce
- positive safety promotion policy

Also from the literature, Donald and Canter identified eight additional factors, derived using expert judgement, that discriminated between factories (locations) in terms of safety climate. In order of decreasing discriminant power they were:

- importance of safety training
- effects of workpace [sic]
- status of safety committee

- status of safety officer
- effect of safe conduct on promotion
- level of risk at the workplace
- management attitudes toward safety
- effect of safe conduct on social status

These factors are summarized in the first column of Table 1.

**Table 1**
**Attributes Related to Safety in the Chemical Industry [24]**

| Attributes derived from the literature: | Proposed attributes to be tested empirically: |
|---|---|
| • management commitment<br>• safety training<br>• open communications<br>• environmental control and management<br>• stable workforce<br>• positive safety promotion policy | *People facet*<br>• self<br>• workmate<br>• manager<br>• supervisor<br>• safety representatives* |
| **Attributes found using expert judgement:**<br>• importance of safety training<br>• effects of workpace (*sic*)<br>• status of safety committee<br>• status of safety officer<br>• effect of safe conduct on promotion<br>• level of risk at the workplace<br>• management attitudes towards safety<br>• effect of safe conduct on social status | *Attitude behavior facet*<br>• satisfaction<br>• knowledge<br>• action<br><br>*Activity facet*<br>• active<br>• passive |
|  | _____<br>* Attribute marked with an asterisk did <u>not</u> correlate with low self reported accident rates. |

As their evaluation technique, Donald and Canter used an employee survey based on three facets of safety attitude: (1) people, (2) attitude behavior, and (3) activity. The "people" facet was divided into five components: (1) self, (2) workmates, (3) supervisors, (4) managers, and (5) safety representatives. The attitude behavior facet was divided into three components. The first was an employee "knowing about" something related to safety, the second was an employee being "satisfied with" something about safety, and the third was an employee "carrying out" some action related to safety. Finally the "activity" facet addressed the degree to which an employee engaged in activities that are important to safety. The elements of each of the three facets were used to construct ten "scales" to measure worker attitudes toward safety and their perception of other people's attitudes. The ten scales are summarized in the second column of Table 1.

Donald and Canter then designed question templates to map all three facets into specific questions related to safety climate. One such template, the combination of "workmate (people facet) is satisfied with (attitude facet) passive safety activity (activity facet)," would lead to the development of questions like, "To what extent are your workmates satisfied with the safety procedures they are required to follow?"

Thus, Donald and Canter represented each of the ten scales by a set of questions developed from the templates described above. In addition, they asked each participant in the survey about their involvement in accidents. These "self reported accident rates" were the safety metric chosen for the study.

Donald and Canter conducted the survey at ten plants owned by the same company. The results indicated that the attitude scales were a reliable measure of safety climate. Only one scale, safety representatives, did not show a statistically significant correlation with self-reported accident rates. Overall, there was a "...clear and strong relationship between the safety attitude climate of a company and its accident performance."

The study discussed in the preceding paragraphs can be summarized in terms of the framework displayed in Figure 2. The ten scales representing the attributes of safety climate (second column of Table 1) are the proposed safety culture attributes resulting from step 1. Step 2, the design of evaluation techniques, is the design of the question templates and question sets. Data collection and analysis, step 3, is administering the questionnaires and verifying the validity of the responses. The external safety metrics needed as input to step 4 were the self-reported accident rates solicited in the questionnaires. Correlating the measured values of each of the 10 scales with the safety metrics was presumably done using regression analysis, although this is not explicitly stated. The results of the correlation identified nine of the ten scales as having a significant correlation with self reported accident rates. Thus the output of step 5 is nine of the ten proposed attributes or scales.

Note that the Donald and Canter study described above did not extend to the activities displayed in Figure 3. There was no attempt at identifying performance indicators as surrogates for either the accident rates or the attributes of safety climate, nor was there any attempt to quantify the level of risk represented by particular values of the safety climate scales.

Building on the work of Donald and Canter, Hurst, et al. [25], developed an audit to quantitatively assess safety management systems in the chemical process industry. The project included analyzing loss-of-containment accidents, modeling the safety management system as a control-and-monitoring loop, and a process risk management audit (PRIMA). The audit covered eight key areas, such as hazard reviews of design, supervision of operations, human factors reviews of maintenance, checking/supervision of maintenance, and human factors reviews of operations. The control-and-monitoring loop addressed five levels in the safety management system: (1) system climate, (2) organization and standards, (3) communications, control, and feedback, (4) human reliability, and (5) containment reliability.

Using PRIMA as a starting point, Hale, et al., proposed modeling safety management in a more direct way [26]. They noted that attempts to address safety management as a part of formal risk assessment involve extrapolating from observations of management functions and performance to hardware or human failure probabilities. This extrapolation, they contended, has not been supported by adequate data. Consequently, they proposed to model all processes important to the safety management system using a structured analysis and design technique (SADT) [27], which provides a link between evaluations (audits) of safety management systems at one end, and quantitative risk analysis at the other. The authors described safety audits as strong in their ability to address safety management and certain cultural issues, but unable to describe how culture affects risk levels. Risk analysis, while obviously strong in establishing risk levels, is also unable to establish the link between culture and risk levels. Thus the authors proposed SADT modeling to represent the management processes linking safety culture to work practices and hardware.

## 7.2 Safety Survey of a Nuclear Fuel Reprocessing Plant

Lee [5] reported on an assessment of safety culture at the Sellafield nuclear reprocessing plant, which can perhaps be considered as belonging both to the nuclear industry and the chemical process industry. Lee observed that the concept of safety culture is not new, and had existed for some years as "safety climate," which in turn was one aspect of a broader "organizational climate." He noted that "...many large organizations are currently finding that their efforts to engineer faults out of the system have worked so well that accident rates have reached a low but unassailable plateau and the only way to continue the improvement is to address the hearts and minds of the management and workers."

Lee's description of the traditional approach to safety reflects the process that has been used within the U.S. regulatory system. "The traditional approach to safety ... has been retrospective, built on precedents. Because it is necessary, it is easy to think it is sufficient. It involves, first, a search for the primary (or "root") cause of a specific accident, a decision on whether the cause was an unsafe act or an unsafe condition, and finally the supposed prevention of a recurrence by devising a regulation if an unsafe act, or a technical solution if an unsafe condition." Although maintaining that this process is necessary, Lee went on to note that it has serious shortcomings. Specifically, "Regulations are proliferated to the point where they become incomprehensible and ... resources are diverted to prevent the accident that has happened rather than the one most likely to happen."

Lee observed that "There has been little direct research on the organizational factors that make for a good safety culture. However, there is an extensive literature if we make the *indirect* assumption that a relatively low accident plant must have a relatively good safety culture" (emphasis in original). Lee started with a list of characteristics of plants with low accident rates distilled from a review of empirical research into the organizational aspects of safety. The list included a high level of communication; good organizational learning; a strong focus on safety; strong commitment to safety by senior management; a management leadership style that is democratic, cooperative, participative and humane; more and better quality training; good working conditions; high job satisfaction; and a workforce retained for safe working habits.

These characteristics, summarized in the first column of Table 2, are similar to those used as a starting point by Donald and Canter (the first column of Table 1). Lee then identified 19 attitudes toward safety (safety culture attributes) to be tested empirically. Lee's attributes, listed in the second column of Table 2, also bear some similarity to those examined by Donald and Canter. The evaluation process involved both focus groups and an employee questionnaire consisting of 172 statements about safety. Respondents could indicate a range of agreement or disagreement on a seven-point scale. The safety metric chosen was self reported rates of accidents involving three or more days of lost work.

Lee's results showed a strong correlation between positive safety attitudes and low accident rates. Of the 19 factors, 16 showed a statistically significant correlation, 15 of those at a very high level of significance.

Lee concluded that "The concept of safety culture ... now has widespread support. If it is a valid concept ... [it] should be helpful in getting employees to understand the objectives of a safety management system ... However, the sheer multiplicity of constituent elements of a safety culture and its precept of universal involvement imply that any attempt to monitor its health ... is bound to be complex ..."

## 7.3 An Organizational Analysis Approach

There appear to be three aspects of organizations that can provide the basis for examining a relationship between management and organizational factors (including safety culture) and safety of operations. Specifically, those aspects are (1) the structure of the organization, (2) the processes the organization uses, and (3) the behavior and attitudes of the organization's employees and management personnel. Individual studies have used one, two or all three of these aspects. The studies of the chemical process industry described above focused on employee attitudes. Work begun for the NRC at Pacific Northwest Laboratory (PNL) in the early 1980s [28, 29, 30, 31] focused primarily on the relationship between the structure of the utility organization and safety performance. The first of these reports addressed identifying appropriate organizational factors (step 1 of Figure 2) and possible external safety metrics (the input to step 4).

25

**Table 2**
**Organizational Factors Related to Safety [5]**

| Characteristics of plants with low accident rates: | Proposed safety attitudes (safety culture attributes) to be tested empirically: |
|---|---|
| • high level of communication<br>• good organizational learning<br>• strong focus on safety<br>• strong senior management commitment to safety<br>• democratic, cooperative, humanistic management leadership style<br>• more and better quality training<br>• clean, comfortable working conditions<br>• high job satisfaction<br>• workforce retention is related to working safely | • confidence in safety procedures<br>• personal caution over risks<br>• perceived level of risk at work<br>• trust in workforce<br>• confidence in efficiency of "permit to work" system*<br>• general support for "permit to work" system<br>• perceived need for "permit to work" system*<br>• personal interest in job<br>• contentment with job<br>• satisfaction with work relationships<br>• satisfaction with rewards for good work<br>• personal understanding of safety rules<br>• perceived clarity of safety rules*<br>• satisfaction with training<br>• satisfaction with staff suitability<br>• perceived source of safety suggestions<br>• perceived source of safety actions<br>• perceived personal control over safety<br>• satisfaction with design of plant<br><br>* Attributes marked with an asterisk did <u>not</u> correlate with low accident rates. |

Drawing on work done in organizational analysis, Osborn, et al. [28], proposed a model based on categories of variables they called "organizational contingencies" and "intermediate outcomes." Under the heading of organizational contingencies, potential important organizational factors were grouped into four types: (1) environment, (2) context, (3) governance, and (4) design. The utility environment includes general economic trends, regulation by the State, regulation by the NRC, support from vendor organizations, and interfaces with corporate parents. The second factor, the utility's context, includes its history, size, and technology. Large utilities with a history of introducing new technology may behave differently than smaller utilities that have been technologically conservative. The third factor, organizational governance or management philosophy, is characterized as (1) traditional, which emphasizes a bureaucratic approach including administrative control, written policies, and elaborate written procedures; (2) modern, which emphasizes values where individual judgement is to be used to implement policy; or (3) federal, which stresses negotiation and integration of differing views through conflict resolution. The fourth factor, organizational design, addresses how work is divided among units; the nature of controls placed on individuals, managers, and operating units; coordination mechanisms; and developmental mechanisms that reinforce and direct decisions by individuals.

The second category of variables, called "intermediate outcomes," includes four factors: (1) compliance, (2) efficiency, (3) quality, and (4) innovation. These factors appear to be included in the model to account for organizational characteristics that are closely related to safety and to external regulation. The authors noted that, "Long-term safety appears to be enhanced to the extent a utility promotes quality, compliance, efficiency, innovation, and employee maintenance." These organizational factors (the output of step 1 in Figure 2) are summarized in the first column of Table 3. There was no provision in the model for representing organizational culture or safety culture, except to the extent that it might be reflected in the "organizational governance" factor. The focus was on organizational structure rather than organizational process.

The safety metrics chosen for the PNL study were typical of the early attempts to identify such indicators. The preliminary list included licensee event reports, inspection and enforcement data, operating and outage data, SALP scores, personnel exposure, and operator exam scores.

The work started at PNL continued over a period of about ten years. The objective, of course, was to find statistically valid relationships among the organizational factors, the safety performance indicators, and safe plant operation. Success would involve identifying leading indicators of safety-related performance by examining appropriate organizational factors.

A second report [29], published about a year after the preliminary work, claimed some success with the proposed approach, although the results were still labeled as preliminary. Specifically, the report concluded that (1) plant performance data could be used to create reliable indicators of plant safety performance, (2) plant safety performance indicators are potentially useful for identifying causes of poor performance, and (3) organizational structure appears to be an important predictor of plant safety performance. The organizational structure considered in this case included vertical measures (the number of levels of management), horizontal measures (number of organizational units at each level), and coordination measures (the management level at which operations, maintenance and engineering converge).

The safety metrics chosen were regulatory compliance, hardware failure, human error, and plant reliability. As the authors acknowledge, these elements are performance indicators, rather than direct measures of operational safety. The results of the regression analysis can best be described as mixed. One conclusion, for example, was that facilities with more levels of

**Table 3**
**Management and Organizational Factors**
**Related to Safety Performance**

| **Organizational Analysis Approach [30]** | **Organizational Process Approach [37]** |
|---|---|
| *Environmental Conditions*<br>    General Environment<br>        Abundance of resources<br>        Amount of volatility<br>        Amount of interdependence<br>    Task Environment<br>        Abundance of resources<br>        Amount of volatility<br>        Amount of interdependence | *Administrative Knowledge*<br>    Coordination of Work<br>    Formalization<br>    Organizational Knowledge<br>    Roles and responsibilities |
| *Contextual Conditions*<br>    Size (staff and budget)<br>    Technological<br>        sophistication<br>    Technological variability | *Communications*<br>    External<br>    Interdepartmental<br>    Intradepartmental |
| *Organizational Governance*<br>    Traditional, Modern or<br>        Federal | *Culture*<br>    Organizational Culture<br>    Ownership<br>    Safety Culture<br>    Time urgency |
| *Organizational Design*<br>    Mechanistic, Organic or<br>        Diverse | *Decision Making*<br>    Centralization<br>    Goal Prioritization<br>    Organizational Learning<br>    Problem Identification<br>    Resource Allocation |
| *Intermediate Outcomes*<br>    Efficiency<br>    Compliance<br>    Quality<br>    Innovation | *Human Resource Allocation*<br>    Performance Evaluation<br>    Personnel Selection<br>    Technical Knowledge<br>    Training |

management tend to have more hardware failures and lower plant reliability than facilities with fewer levels of management. Regulatory compliance, however, appeared to be improved by additional levels of management. The reasons for this anomaly were not clear, but the authors also concluded that factors that promote regulatory compliance could either improve or degrade hardware failure rates, human error, and plant reliability. Some results were counterintuitive. No significant correlation was found, for example, between reactor operator examination scores (presumed to be a measure of training effectiveness) and the number of LERs resulting from personnel errors.

Later work by some of the same investigators [30] proposed a framework to link organizational factors and nuclear power plant performance. The framework supports the notion that nuclear power plants are complex entities influenced by internal and external forces, only some of which are under management control. A subsequent report [31] developed the theory required to show how the organizational factors presented in the framework combine to influence nuclear power plant performance. The authors concluded that "Central concepts in the theory of safety in nuclear power plants which have roots in economic and behavioral theories of organizations, are effective in predicting safety-related performance in plants."

The approach described above focused on the structure of the organization. It is based on a body of work in organizational analysis that appears to have virtually no overlap with the proponents of corporate culture. The organizational factors considered are not components of either organizational culture or safety culture, and have different properties. It is interesting to note, however, that the later work by these investigators acknowledges the possible importance of the concept of organizational culture. Specifically, work done following the Bhopal, Challenger, and Chernobyl accidents prompted the authors to note, "Collectively, these analyses suggest that relationships that emerge from the day-to-day operation of technologies are potentially as important as the more general state conditions and management philosophy concerns described earlier. [T]hese management relationships ... are those unplanned continuing dynamics of the organization that allow it to operate with continuity and react to unanticipated conditions. They arise because individuals shape and mold the formal organization, interpret the environment and context, implement the management philosophy and generally add variety to that planned into the system" [30, p. 51].

## 7.4 An Organizational Process Approach

The approach proposed and developed by Haber, et al. [32], is based on organizational processes, as opposed to organizational structure. As with the organizational structure approach pursued at PNL, the underlying idea is to seek statistically valid relationships between organizational factors and safe plant operations. The three-step process used was to (1) develop a description of the human organization of a nuclear power plant, (2) identify organizational and management functions and processes related to safety performance, and (3) develop methods to measure organizational and management factors. The overall concept was designated nuclear organization and management analysis concept (NOMAC).

The development of organizational factors was based on the work of Mintzberg [33], who characterizes organizational structures according to five functional components of the organization and the mechanisms for coordinating work among them. In Mintzberg's model, the five functional components are the operating core (the people who do the work that is central to the organization), the strategic apex (chief executive officers and plant managers), the middle line managers, the technostructure (accountants, trainers, and engineers) and support staff (cafeteria, payroll, security). Within this general structure, Mintzberg identifies five typical structures. Nuclear power plants seem to be best characterized within these structural types as "machine bureaucracies." These are typified by large operating units, personnel grouped by function, centralized decisionmaking, and a sharp distinction between staff and line personnel. One of the most important characteristics of a machine bureaucracy is the standardization of work.

The assessment of management and organizational factors involved three types of data collection: (1) a functional analysis; (2) a behavioral observation technique; and (3) an organizational culture assessment. The functional analysis provides a description of the work flow, behavioral observation identifies patterns of communication, and the culture assessment describes the environment of the organization. Two demonstration studies, one at a fossil power plant and one at a nuclear power plant, identified five organizational factors for further investigation. Specifically, those factors were communication; organizational culture; decision-making; standardization of work processes; and management attention, involvement, and oversight.

In this work, organizational culture was described as "... the beliefs, perceptions, and expectations that individuals have about the organization in which they work and about the values and consequences that will follow from one course of action or another. Consequently, culture highly influences behavior within the organization." Safety culture was considered to be an element of organizational culture. Organizational culture in the demonstration studies was measured using organizational culture assessment questionnaires distributed to operations, maintenance, engineering and support employees. Each questionnaire covered six topics: (1) organizational culture inventory; (2) cohesiveness; (3) commitment; (4) hazardous nature of work; (5) safety; and (6) routinization.

The organizational culture inventory covered 12 factors representing thinking and behavioral styles that might be expected from members of the organization. The factors included such characteristics as humanistic (the organization is managed in a person-centered way), approval (conflicts are avoided and personal relations are pleasant), oppositional (confrontation prevails and negativism is rewarded), and perfectionistic (persistence, hard work, and perfectionism are valued).

The other five topics in the assessment measured different aspects of organizational culture:

- Cohesiveness - a measure of the relative strength of an individual's identification with and involvement in a particular work group

- Commitment - a measure of an individual's identification with and involvement in a particular organization

- Hazard - a measure of an individual's perception of the hazardous nature of their work

- Safety - a measure of an individual's perception of the importance of safety to success in the organization

- Routinization - a measure of the amount of repetition people perceive in their work on a day to day basis

A subsequent paper by Haber, et al. [34], included data collected from a second nuclear power plant. Both nuclear plants were chosen "because of their excellent safety records and because they are considered "good performers" by the NRC." An example of the hypotheses tested in this approach is, "A higher value on the communication organizational variable would result in a lower total number of human error LERs" [35]. The survey data collected during the organizational culture assessments showed both similarities and differences between the two. Actual performance indicators were not reported for either plant.

Jacobs, et al. [36], adopted a similar viewpoint based on organizational processes. They also suggested that multiple assessment methods were required to meet the research goals of identifying and assessing organizational factors related to nuclear power plant safety. Specifically, they formulated three assessment packages to be used to conduct interviews of plant personnel. One module was directed at upper management, one at departmental processes, and one at interdepartmental relationships. A subsequent paper [13], identified five organizational factors as relevant to safe operations: culture, administrative knowledge, communications, decisionmaking, and human resource allocation. This is similar to the list developed by Haber, et al. Haber's "standardization of work processes" and "management attention" are omitted, and "administrative knowledge" and "human resource allocation" are included. In addition to identifying the five factors, Jacobs assigned several dimensions to each, as shown in the second column in Table 3 of this report. A somewhat later joint paper by Jacobs and Haber [37] used Jacob's list of organizational factors and dimensions.

Organizational culture, as used by Jacobs, et al., has somewhat different dimensions than the same term used by Haber. Specifically, Jacobs, et al., defined organizational culture as plant personnel's shared perceptions of the organization, including traditions values, customs, practices, goals, and socialization. They also treated it as a subdivision of a broader factor called "culture." The general factor "culture" was subdivided into organizational culture, ownership, safety culture and time urgency. Ownership includes the degree to which personnel take responsibility for their actions. Time urgency is the degree to which plant personnel feel schedule pressures. Safety culture is described as the characteristics of the work environment, such as common understandings, that influence plant personnel's perceptions of the importance that the organization places on safety. It includes the degree to which a critical, questioning attitude exists. This structure, and the relationship among culture, organizational culture and safety culture, are not so much at odds with Haber, et al., as they are simply different.

The work reported by Haber, et al. [34, 35], and Jacobs, et al. [36, 37], concentrated primarily on defining the management and organizational factors (or safety culture attributes) that could influence safety of operations, and designing the formal evaluation techniques (steps 1 and 2 in Figure 2 of this report). Data collection and analysis (step 3) appears to have been limited to one fossil-fired power plant and two nuclear power plants. Work reported on correlating data with safety metrics (step 4) and on using management and organizational factors to calculate new risk metrics (Figure 3) appears to be limited to a few examples.

Table 3 lists the organizational factors chosen for investigation in the organizational analysis described earlier [30], and those proposed by Jacobs and Haber [37]. A comparison of the two columns shows the emphasis on structure and conditions in the first column and the emphasis on process in the second column. Both approaches are designed to relate management and organizational factors to safety performance. The differences illustrate the disparate viewpoints that can be brought to the same undertaking. One difference between the two perspectives that is not immediately obvious, is that the organizational analysis approach relies primarily on data that can be obtained from organizational documents and records. In fact, the approach was designed to rely as much as possible on publicly available records. The organizational process approach, on the other hand, depends heavily on inferring organizational characteristics from surveys and interviews of a broad spectrum of personnel in the organization. It attempts to determine how an organization really works, rather than how it is structured. Characterizing organizational culture, in particular, depends on survey data. The collection of such data is labor intensive and can be intrusive.

7.5 Work Process Analysis

In 1994, Davoudian, Wu and Apostolakis [38, 39] proposed an approach to modeling the organization that uses elements of both organizational structure and organizational process. The ultimate goal is to develop a methodology for incorporating organizational factors into probabilistic risk assessments. As reported in several papers [16, 40, 41, 42], this work has evolved over the past few years. The analysis begins with asking the question "how is the organization supposed to work?" and then addressing "how well is it working?". This work adopts the categories and dimensions of important organizational factors as articulated by Jacobs and Haber [37]. It then proposes an examination of work processes as a way to analyze and possibly quantify the importance of those factors.

The work process analysis methodology begins with the observation that the structure of an organization is determined by two basic elements, namely division of labor and coordination of effort. Division of labor is accomplished by creating work units, which are typically based on functional specialization. Examples are operations, maintenance, instrumentation and control, and health physics. Coordination is accomplished by both formal and informal mechanisms, including policies, procedures, scheduled meetings and unscheduled meetings. Work processes within a functional unit tend to be standardized and controlled by written procedures. The objective of the work process analysis methodology is to identify the organizational factors that then can impact the performance of particular tasks, and ultimately to quantify those impacts as changes in PRA parameters (failure rates, human error probabilities, or system unavailabilities).

The first step in the work process analysis model (WPAM) is the identification of front-line and supporting work processes. Front-line processes are those that have a direct influence on the operability of plant hardware, such as plant operation, maintenance, and modifications. Supporting work processes include training, procurement, and quality control. For each work process, WPAM poses the basic question "How can an accumulation of organization failures lead to an unsafe plant condition?". The analysis of each task leads to a flow diagram that displays each major step in the process, along with the defenses or barriers established to ensure correct execution. A flow diagram for a corrective maintenance work process is shown in Figure 4.

Each task in a given work process can be influenced by several organizational factors. In fact, one of the strengths suggested for this approach is its ability to identify organizational deficiencies that could disable dissimilar components. If the analysis is to be extended to quantifying the impacts on human error rates or system unavailability, it is necessary to rank the organizational factors according to their degree of influence on each task. One method of performing this ranking is the analytical hierarchy process (AHP), which involves assigning relative weights to each pair of pertinent factors (pairwise comparison). Presumably, other ranking methods could be used.

In a 1999 paper [43], Weil and Apostolakis proposed that the 20 organizational factors identified by Jacobs & Haber [37] can be reduced to six without impairing the effectiveness of the work process analysis methodology. The six factors retained are communications, formalization, goal prioritization, problem identification, roles and responsibilities, and technical knowledge. These six were chosen by identifying factors that affect a large number of tasks and/or are often cited as contributing to errors, and by excluding factors that logically could be combined into one of the remaining factors.

In terms of the framework represented in Figure 2, WPAM primarily addresses step 2, design of evaluation techniques, and step 3, data collection and analysis. It can perhaps best be thought of as a method for root cause determinations. This methodology can also be extended to include quantifying the effects of organizational processes, and the calculation of changes in risk metrics [39], as indicated by the lower path in Figure 3.

# Figure 4

## Flow Diagram for Corrective Maintenance Work Process

(from Weil & Apostolakis [43])

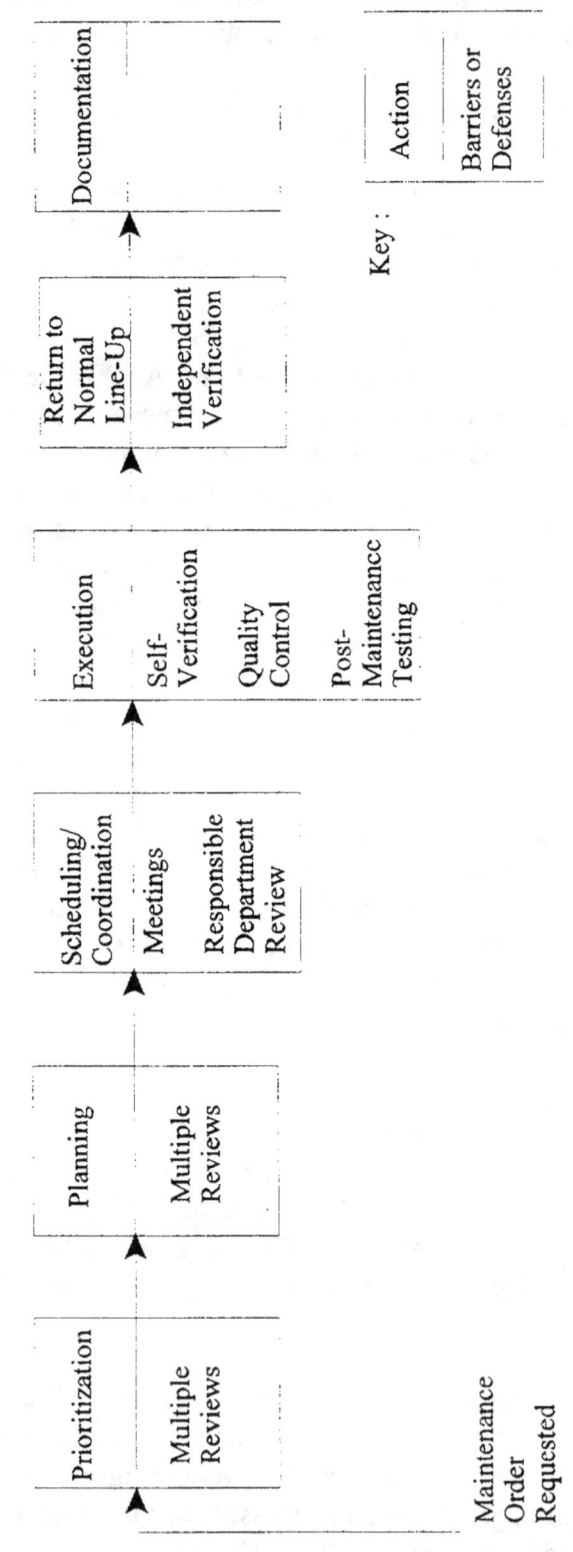

[Reprinted from *Reliability Engineering and System Safety*, Volume 45, Nos. 1-2: "Probabilistic Safety Assessment Methodology," G.E. Apostolakis, et al, pp. 85–105, 1994, with permission from Elsevier Science. ]

## 7.6 A Model Based on Expert Elicitation

The Swedish Nuclear Power Inspectorate (SKi) sponsored a study to develop a risk-based performance monitoring system for nuclear power plants using expert elicitation to identify organizational and operational-based safety-related performance indicators [44]. This model is based on a probabilistic safety assessment of the plant. Starting with a proposed list of 78 performance indicators, the study derived a final list of five high worth indicators:

- annual rate of safety significant errors

- annual rate of maintenance problems

- ratio of corrective to preventive maintenance on safety equipment

- annual rate of problems with repeated root cause

- annual rate of plant changes that are not incorporated into design-basis documents prior to the next outage

The SKi proposed these five indicators as a suitable measure of safety culture. Ultimately, the assessment of safety culture (superior, above average, average, below average, or inferior) can be used to modify equipment failure rates or system unavailabilities.

The SKi process is particularly interesting because it replaces virtually all of the activities represented in Figure 2 with expert elicitation. Step 1 is represented by the initially proposed list of 78 performance indicators. Steps 2 through 5, are replaced by the expert elicitation process, with the output being the final list of five high-worth indicators. Since the methodology includes an algorithm for quantifying the impact of the performance indicators on risk metrics, it provides a means of addressing the upper path in Figure 3.

## 7.7 A Summary of the Empirical Evidence

There is a substantial body of literature dealing with the relationship between safety culture and safety of operations. That literature is fragmented, however, and it is often difficult to understand how one piece of work relates to another, if at all. The scope, depth, terminology, and perspective vary widely from one study to the next.

The first source of difficulty is terminology. There is general agreement on the concept of safety culture, and some agreement on its attributes. Many of the studies that relate management and organizational factors to safety of operations do not use the term "safety culture." If it is used, it may denote a narrowly defined element of a larger set of management and organizational factors being investigated. One study can only be compared with another by looking at the organizational attributes that are actually measured. The study of safety culture might benefit substantially if a consensus could be developed on its definition and, most importantly, its measurable attributes.

A second source of difficulty is the availability of suitable safety metrics. The chemical processing and transportation industries have sufficiently high occurrences of unwanted events that it is possible to correlate management and organizational factors with accident rates. Other activities, including nuclear power generation, have sufficiently low accident rates that they provide no basis for comparing one facility to another. Instead, investigators select performance indicators, such as the number of unplanned scrams, as surrogates for safety performance. Some investigators believe that the validity of using performance indicators requires more justification than it has received.

Olson, et al. [45], illustrate the issue by distinguishing among three categories of information: (1) plant performance indicators, (2) penultimate measures of safety, and (3) ultimate measures of plant safety. They suggested that the ultimate measures of plant safety are the unwanted events, such as core melt, large releases of radionuclides, and large population exposures. The penultimate measures of safety are potentially significant events, releases of radionuclides, and personnel exposures. Analyses of potentially significant events to determine conditional core damage probability or conditional large early release frequency can partially bridge the gap between the penultimate and ultimate measures of safety.

Plant performance indicators might include the number of LERs, operating and outage data, and the number of violations of NRC regulations. The use of performance indicators as measures of safety should include establishing a relationship between the indicator and the likelihood of an unwanted event. Current NRC work to identify risk-based performance indicators is intended to address this issue [46].

No studies relating safety culture and safety of operations were identified which addressed all of the activities outlined in Figures 2 and 3. Studies of the chemical process industry addressed all of the activities in Figure 2, and provided empirical evidence that safety attitudes have a positive relationship to safety of operations. Those studies did not address identifying performance indicators.

Studies of nuclear power plants focused on identifying management and organizational factors that are important to safety of operations, but they lack the extensive field data collected in the chemical process industry studies. The work started at PNL by Osborn, et al. [28], involved extensive empirical analyses relating organizational factors to performance indicators, but did not examine attributes of safety culture. The work begun at Brookhaven National Laboratory (BNL) by Haber, et al. [32], did address organizational factors related to safety culture, but data collection and analysis concentrated on measuring those attributes and validating the measurements. Data collection appears to have been limited to one fossil and two nuclear power plants, and very little was reported on establishing an empirical relationship between the organizational factors and indicators of safety performance.

Overall, substantial work has been done to validate the idea that safety culture and other management and organizational factors have a strong relationship to safety of operations. Most of the empirical work has been done outside the nuclear industry. Some investigators believe that results cannot be extrapolated from one industry to another without justification that does not now exist [26]. It appears that Lee [5] has characterized the situation correctly, in saying that "There has been little direct research on the organizational factors that make for a good safety culture. However, there is an extensive literature if we make the indirect assumption that a relatively low accident plant must have a relatively good safety culture." The proponents of safety culture as a determinant of operational safety in the nuclear power industry rely, at least to some degree, on that indirect assumption.

## 8. REGULATORY PERSPECTIVES

Regulatory organizations have an interest in safety culture because it is now widely believed that there is a relationship between safety culture and safety of operations. The most obvious link suggested by work done to date is that a good safety culture is expected to reduce human error rates. Reason [12] suggests that well-defended technologies (meaning those technologies that make extensive use of defense in depth, such as nuclear power) may be especially vulnerable to an unsafe culture. He points out that the effect of a poor safety culture is to create gaps or holes in the defenses which are not readily apparent (latent errors), thus making the system vulnerable to a serious accident when the right initiating event occurs. Defenses in depth make the system more opaque to the operators, and the operators are more remote from the processes they control. An important question remains as to whether the regulatory process should address safety culture. That question probably cannot be answered without considering *how* the regulatory process might address safety culture.

INSAG [7] asserts that safety culture is attitudinal as well as structural, and that it relates to both organizations and individuals. Lee [5] suggests that the safety improvements to be achieved through engineering are limited, and that additional improvements require addressing the "hearts and minds of the management and workers." Reason [12] speculates that safety culture accounts for accident rates varying by a factor of 40 from the best to the worst of the commercial air carriers. Studies sponsored by the NRC [31 and 47] have shown a positive correlation between management and organizational factors and selected safety indicators. Studies outside of the nuclear industry [24 and 25] have shown strong positive correlations between organizational characteristics associated with safety culture and low accident rates.

Although there is no universally accepted definition, there is some common ground among investigators on the elements of safety culture. Most investigators appear to agree that the elements include good communication, organizational learning, senior management commitment to safety, and a working environment that rewards identifying safety issues. Some investigators would also include management and organizational factors such as a participative management leadership style. The regulatory dilemma is that the elements that are important to safety culture are difficult, if not impossible, to separate from the management of the organization.

Historically, the NRC has been reluctant to regulate management functions in any direct way. Licensees have been even more reluctant to permit any moves in that direction. The argument is, of course, that licensees are responsible for the safe operation of their facilities, and they must be permitted to achieve safety in their own operating environment in the best ways they know. The closest the NRC has come to evaluating management performance is the systematic assessment of licensee performance (SALP) program, which the agency has discontinued. Throughout its life, licensees criticized the SALP program as lacking objectivity.

Former NRC Commissioner Kenneth Rogers [48] suggested that there are three essential regulatory functions. The first is to ensure that the facility operator accepts the responsibility for safe operations, and the regulator does nothing to diminish it. The second is for the regulator to assure itself that the facility operator is achieving and maintaining the desired level of safety. The third function is to provide assurance to the public that the regulatory agency is doing its job. Viewed in this way, it becomes clear that the regulator's function is not to step in and provide safety if the licensee falls short. Rather, the regulator's activities should be designed to evaluate the relationship between the licensee's activities and safety, and to warn the licensee when that evaluation indicates a decline in safety. It can be argued that examining a licensee's activity and evaluating its effectiveness need not lead to telling the licensee how the activity should be conducted. It would seem to follow that, to the extent the NRC does not now understand how safety culture affects safety of operations, and how the regulatory process affects safety culture, there is additional research to be done.

8.1 Safety Culture as a Basis for Safety Regulation

One of the most comprehensive reviews of the relationship between safety culture and safety of operations was undertaken for the United Kingdom Health and Safety Executive (HSE) by its Advisory Committee for the Safety of Nuclear Installations (ACSNI).

As defined by INSAG, "safety culture" is idealized. The construct does not explicitly treat the existence of an inadequate safety culture that can be improved. The ACSNI suggested a different working definition: "The safety culture of an organization is the product of individual and group values, attitudes, perceptions, competencies, and patterns of behavior that determine the commitment to, and the style and proficiency of, an organization's health and safety management. Organizations with a positive safety culture are characterized by communications founded on mutual trust, by shared perceptions of the importance of safety and by confidence in the efficacy of preventive measures" [49].

On the basis of work sponsored by the NRC, ACSNI concluded that the key predictive indicators of safety performance are effective communication, good organizational learning, organizational focus on safety, and external factors such as the financial health of the organization and the impact of regulatory bodies. ACSNI holds that "The best safety standards can arguably only be achieved by a programme which has a scope well beyond the traditional pattern of safety management functions." ACSNI characterizes the evolution of safety regulation as follows:

"There are three phases in the history of attempts to regulate general industrial safety.

"First, there is a stage of concentration on the outcome; if a worker or a member of the public is harmed, those considered responsible are punished.

"Second, there is a stage of prescribing in advance the detailed action that industry must take. For example the organisation must provide guards of certain types for specific machines ... This stage is an advance because it attacks points of danger before actual harm occurs ...

"In the third stage, industry is canvassed to develop a "safety culture". ... This stage of regulation ... concentrates on the internal climate and organization of the system [and] also emphasizes the need for every individual to "own" the actions being taken to improve safety ... "

In examining the regulator's role in influencing licensees' organizational behavior, the ACSNI human factors study group suggested that, "If organizational failure is as important as, say, the physical integrity of the plant, it must be considered with equal care. This implies that the qualifications and skills of the regulator must include this topic as well. In particular, three areas are considered important. Firstly, the need for a sufficient number of inspectors to have appropriate formal qualifications. Secondly, a need generally for adequate training in interpersonal skills, to ensure that the regulator ... does not give the impression of being prescriptive. Thirdly, the need for the regulator when visiting licensed sites... to be aware of the general indicators which reflect the state of the licensee's safety culture." Furthermore, "The behavior of the regulators will affect the culture of the licensees ... The regulators need to act in such a way as to encourage "ownership" of safety by the whole staff of the licensee" [49, p. 47].

A theme that runs through the ACSNI study is that the most effective safety cultures will develop in less-prescriptive regulatory structures. "The most impressive achievements appear in companies where the pressure for safety has been generated from within the organization, apparently independent of external standards" [49, p.16]. This theme is reinforced in a subsequent report for HSE by Four Elements, Ltd., which noted that "The form of safety culture cannot be regulated ... The form of regulation can influence safety culture positively or negatively ..." [50, p. 10].

The Four Elements report also noted that "It is recognized that there are a number of prescriptive regimes, such as the U.S. Nuclear Industry, where the encouragement of a positive safety culture is still essential. It is considered that those Operators with good Safety Cultures, within the US regulatory regime, tend to self-regulate around the constraints of the regulatory regime, to attain levels of safety which are beyond those minima specified in the regulations. The manner in which the Regulator can encourage such self regulation is not clear [50, p. 34]."

One aspect of this idea is explored in some detail in an earlier paper by Marcus [51], which examined the implementation of certain NRC requirements at several U.S. nuclear power plants. His conclusion was that "... nuclear power plants with relatively poor safety records tended to respond in a rule-bound manner that perpetuated their poor safety performance and that nuclear power plants whose safety records were relatively strong tended to retain their autonomy, a response that reinforced their strong safety performance."

8.2 International Activities

The International Atomic Energy Agency has continued to develop the concept of safety culture as an important contributor to safety of operations and, therefore, as an important issue to be addressed by the regulatory process. A 1998 IAEA publication is devoted to offering "... practical advice to assist in the development, improvement or evaluation of a progressive safety culture" [52]. A revision to INSAG-3, issued in 1999, listed the improvements made over the original version, beginning with "A more comprehensive treatment of safety culture and defense in depth" [53]. INSAG-13, "Management of Operational Safety in Nuclear Power Plants," was also issued in 1999 [54]. Its stated purpose is "... to build upon the ideas outlined in 75-INSAG-4 [Safety Culture] and to develop a set of universal features for an effective safety management system in order to develop a common understanding."

The Organization for Economic Cooperation and Development (OECD) Nuclear Energy Agency (NEA) has also become engaged in promoting safety culture as an important part of safety regulation. A 1999 NEA publication, "The Role of the Nuclear Regulator in Promoting and Evaluating Safety Culture" [55], suggests signs that a regulator should look for to determine the strength of a licensee's safety culture. It also provides suggestions for regulatory response to a weakening safety culture, although the suggestions are very general. A subsequent NEA report, published in 2000, deals more specifically with the issue of regulatory response [56].

NEA also issued a "state-of-the-art" report [57] on the identification and assessment of organizational factors related to nuclear power plant safety. Volume 1 lists and discusses 12 organizational factors:

- external influences
- goals and strategies
- management functions and overview
- resource allocation
- human resources management
- training
- co-ordination of work
- organizational knowledge
- proceduralization
- organizational culture
- organizational learning
- communication

NEA describes each of these factors in terms of between five and 20 aspects. This list is similar, but not identical, to the list of attributes proposed by Jacobs and Haber (Table 3, column 2).

The second volume of the NEA report summarizes the regulatory framework used in nine OECD countries, including France, the United Kingdom and the United States. In each case, the discussion addresses how the regulatory process considers management and organizational factors. Most of the regulatory programs discussed include some evaluation of management and organizational factors. By contrast, the NRC program does not involve direct evaluation of management performance. Rather, the NRC " ... infers licensee organization management performance based on a comprehensive review of inspection findings, licensee amendments, event reports, enforcement history, and performance indicators." At the other extreme, the NEA report describes the Canadian regulatory program as including direct assessment of 19 of the 20 management and organizational factors developed by Jacobs and Haber [37].

Volume 2 of the NEA report [57] also provides summaries of research programs on management and organizational factors. The programs described are directed at identifying management and organizational factors important to safety of operations, incorporating management and organization factors into PRAs, or evaluating the attributes of safety culture within a licensee's organization. The research programs ascribed to the NRC include the BNL work by Haber, et al. [32] (described previously in this report), the Human Performance Investigation Process [58], and the work on ATHEANA [20].

8.3 Safety Culture and NRC's Regulatory Process

The Commission's "Policy Statement On the Conduct of Nuclear Power Plant Operations" [59], makes safety culture part of the NRC's regulatory agenda. Issued in 1989, the policy statement includes the provision that "Management has a duty and obligation to foster the development of a "safety culture" at each facility and to provide a professional working environment, in the control room and throughout the facility, that assures safe operations." The policy statement then defines "safety culture" using the definition from INSAG-4.

A second Commission policy statement [60] addresses an issue related to safety culture. Sometimes referred to as the policy on maintaining a safety conscious work environment, its actual title is "Freedom of Employees in the Nuclear Industry to Raise Safety Concerns Without Fear of Retaliation." The policy statement, published in 1996, focuses fairly narrowly on the issue of employees' ability to raise safety issues and uses the term "safety conscious work environment" rather than "safety culture" to describe the conditions that the policy is intended to promote.

Inspection Procedure 71841, "Human Performance" [61], of the NRC Inspection Manual, does not use the term "safety culture," but the topic areas and elements to be inspected include many of the attributes or manifestations of safety culture identified in this report. Inspection reports sometimes use the term in describing inspection findings, but often in response to a licensee's use of the term. A report of an inspection at VECTRA Technologies, for example, criticized a management index used to monitor safety culture because it did not provide a "continuous ... monitoring of the safety culture, which ... is needed since VECTRA's safety culture ... has not yet matured" [62].

Current NRC programs to develop risk-informed regulatory processes and performance-based reactor oversight appear to be in consonance with the idea of some degree of self-regulation. The reactor oversight program [63] identifies a level of performance, as measured by a set of performance indicators, where regulatory involvement will be limited to a baseline inspection program. The program identifies seven cornerstones of safety performance, each monitored by one or more performance indicators. The four cornerstones for reactor safety are (1) initiating events, (2) mitigating systems, (3) barrier integrity, and (4) emergency preparedness. In addition to the cornerstones, the staff has identified three "cross-cutting" elements that are part of each cornerstone. These are (1) human performance, (2) management attention to safety and worker's ability to raise safety issues (safety-conscious work environment), and (3) finding and fixing problems (corrective action programs). There are currently no performance indicators associated with these cross-cutting issues.

The NRC staff recognizes that the new oversight program will involve a shift in the NRC's role from improving human reliability to monitoring human reliability. This appears to be consistent with the thought of allowing more of what might be termed "self-regulation." On the other hand, the staff equates the term "safety culture" with "safety conscious work environment." This appears to be a much narrower concept of safety culture than is used by most writers in the organizational safety field.

Two questions are suggested here. The first is whether the NRC is giving sufficient attention to the staff skills, knowledge and abilities that will be required in a risk-informed, performance-based regulatory scheme. If the NRC is to encourage safety culture, it may require a different perspective on the part of the front-line inspection staff. The second question is whether appropriate attention is being given to identifying performance indicators for human performance, safety culture, or other relevant management and organizational factors.

The state-of-the-art on indicators of human performance does not appear to be particularly well advanced. Nishijima [64] has proposed human performance indicators based on three aspects of individual and organization behavior: safety, efficiency, and welfare. For individuals, under safety, he suggested safety consciousness (perhaps measured by a safety attitude index), safety behavior (measured by accident rate), and safety character. Under efficiency, he suggested worker abilities (qualifications and training) and aptitudes. Under welfare, he listed work ethic and sense of belonging. Nishijima also identified parallel organizational factors. His proposal is best characterized as preliminary, and the hard work of relating the proposed indicators to operational safety is left to the future.

The ACSNI study group [49] concluded that research is required particularly in two areas. "Firstly, work is necessary simply to increase the number of validated culture and performance indicators available. Secondly, studies are required to establish the extent to which the indicators remain valid once they have been identified and used as indicators."

The Four Elements report [50] to HSE noted that "... [although] there is much discussion in the literature of Safety Culture, and a widespread recognition of its importance for safety performance, there is little useful information which addresses the mechanism whereby Safety culture affects performance." Neither INSAG-4 nor the ACSNI human factors study group "... clearly identify the link between any one of these attributes [of safety culture] and Safety Culture itself. It is not clear how regulatory or other external influences which affect certain attributes of Safety Culture ... are actually affecting the underlying Safety Culture. ... [Although] both ACSNI and INSAG note the importance of the regulatory influence, neither goes further to describe the mechanisms of that influence" [50].

## 9. CONCLUSIONS

There is a clear consensus among writers in the field of safety management that worker attitudes toward safety make a difference. What is not clear is the mechanism by which attitudes, or safety culture, affect the safety of operations. Statistical evidence that unambiguously links safety culture or its specific attributes with the safety of operations is surprisingly rare, especially within the nuclear industry.

Pidgeon [65] examined the key theoretical issues underlying the concept of safety culture. He noted that "...some 10 years on from Chernobyl, the existing empirical attempts to study safety culture and its relationship to organizational outcomes have remained unsystematic, fragmented, and in particular underspecified in theoretical terms." He identified four theoretical issues that must be addressed if the concept of safety culture is to realize its promise. The first is the paradox that culture can act simultaneously as a precondition for safe operations and an incubator for hazards. The second issue is that in complex and ill-structured risk situations, decisionmakers are faced not only with the matter of risk, but also with fundamental uncertainty characterized by incompleteness of knowledge. The third issue is the organizational construction of acceptable risk. The fourth is the issue of organizational learning and the political need to assign blame for disasters. Pidgeon's paper stressed the importance of safety culture as a concept that is uniquely capable of improving safety in complex systems.

From the narrow perspective of the nuclear power industry, an important next step in understanding the relationship among safety culture, safety of operations and safety regulation would be to develop consensus regarding the essential attributes of safety culture and to identify suitable performance indicators. Consensus may not be easily reached, but investigators seem to have made too little use of past work, and constructed new frameworks rather than building on what has been done. Performance indicators will be even more difficult. Some work is underway to determine the degree to which the performance indicators in the reactor oversight program capture human performance issues. The results of that work might provide some insights into how performance indicators could be developed.

The NRC's regulatory program must ensure that licensees' root-cause analyses and corrective action programs are capable of identifying safety culture issues. Models for human performance, such as ATHEANA [20], will not be realistic until the influence of the plant's safety culture on the "error-forcing context" is assessed [66].

Ultimately, the NRC will have to arrive at an understanding of how its regulatory process can affect the safety cultures of its licensees, both positively and negatively. The role of the regulator needs to be determined, including the possibility that there is no role other than monitoring.

# 10. REFERENCES

## 10.1 References Cited

(1)     International Nuclear Safety Advisory Group, "Summary Report On the Post-Accident Review Meeting On the Chernobyl Accident," Safety Series No. 75-INSAG-1, International Atomic Energy Agency, Vienna, 1986.

(2)     Rogovin, Mitchell, "Three Mile Island – A Report to the Commissioners and the Public," Vol. 1, January 1980.

(3)     Stello, Victor, Letter to NRC Chairman Lando Zech, "Assessing Plant Performance as it Relates to Plant Management," U.S. Nuclear Regulatory Commission, August 17, 1988.

(4)     Ostram, L., C. Wilhelmsen, and B. Kaplan, "Assessing Safety Culture," *Nuclear Safety*, Vol. 34, No. 2, April–June 1993.

(5)     Lee, Terence, "Assessment of Safety Culture at a Nuclear Reprocessing Plant," *Work and Stress*, Volume 12, Number 3, p. 217, Taylor & Francis, Ltd., July–September 1998.

(6)     International Nuclear Safety Advisory Group, "Basic Safety Principles for Nuclear Power Plants," Safety Series No. 75-INSAG-3, International Atomic Energy Agency, Vienna, 1988.

(7)     International Nuclear Safety Advisory Group, "Safety Culture," Safety Series No. 75-INSAG-4, International Atomic Energy Agency, Vienna, 1991.

(8)     International Atomic Energy Agency, "ASCOT Guidelines: Guidelines for Organizational Self-Assessment of Safety Culture and for Reviews by the Assessment of Safety Culture in Organizations Team," IAEA-TECDOC-860, Vienna, 1996.

(9)     Storey, John, *An Introduction to Cultural Theory and Popular Culture*, University of Georgia Press, 1998.

(10)    Deal, Terrence E., and Allen A. Kennedy, *Corporate Cultures: The Rites and Rituals of Corporate Life*, Addison-Wesley, 1982.

(11)    Uttal, Bro, "The Corporate Culture Vultures," *Fortune*, October 17, 1983.

(12)    Reason, James, *Managing the Risks of Organizational Accidents*, Ashgate, 1997.

(13)    Jacobs, Rick, John Mathieu, Frank Landy, Tony Baratta, Gordon Robinson, David Hofmann, and Kathleen Ringenbach, "Organizational Processes and Nuclear Power Plant

Safety," Proceedings of the Probabilistic Safety Assessment International Topical Meeting, Clearwater Beach, Florida, January 26–29, 1993.

(14)     Bridges, William, *The Character of Organizations*, Consulting Psychologists Press, Inc., 1992.

(15)     Merritt, A.C., and R.L. Helmreich, "Creating and Sustaining a Safety Culture," *CRM Advocate*, 1, 8-12, NASA/UT/FAA Aerospace Crew Research Project, Austin, Texas, 1996.

(16)     Apostolakis, G., and J.-S. Wu, "A Structured Approach to the Assessment of the Quality Culture in Nuclear Installations," Presented at the American Nuclear Society International Topical Meeting on Safety Culture in Nuclear Installations, Vienna, April 24–28, 1995.

(17)     Carroll, John S., "Safety Culture as an Ongoing Process: Culture Surveys as Opportunities for Enquiry and Change," *Work and Stress*, Volume 12, Number 3, July–September 1998, p. 272, Taylor & Francis, Ltd.

(18)     Moray, Neville P. and Beverly M. Huey, (eds.), "Human Factors Research and Nuclear Safety," Committee on Human Factors, Commission on Behavioral and Social Sciences and Education, National Research Council, National Academy Press, Washington, D.C., 1988.

(19)     Reason, James, *Human Error*, Cambridge University Press, 1990.

(20)     U.S. Nuclear Regulatory Commission, "Technical Basis and Implementation Guidelines for A Technique for Human Event Analysis (ATHEANA)," NUREG-1624, Rev. 1, May 2000.

(21)     Kaufman, John V. and Sanford L. Israel, "Reactor Coolant System Blowdown at Wolf Creek on September 17, 1994," *Nuclear Safety*, Vol. 36, No. 2, July–December 1995.

(22)     Idaho National Engineering and Environmental Laboratory, Report No. CCN 00-005421, "Summary of INEEL Findings on Human Performance During Operating Events," transmitted by letter dated February 29, 2000.

(23)     Shiel, Tom, "The Human Performance Improvement Program at Duke Power Nuclear Stations," *Nuclear News*, May 2000.

(24)     Donald, Ian, and David Canter, "Employee Attitudes and Safety in the Chemical Industry," *Journal of Loss Prevention in Process Industries*, Butterworth-Heinemann, Ltd., 1994.

(25)     Hurst, Nick W., Stephen Young, Ian Donald, Huw Gibson, and Andre Muyselaar, "Measures of Safety Management Performance and Attitudes to Safety at Major Hazard Sites," *Journal of Loss Prevention in Process Industries*, Vol. 9, No. 2, pp. 161–172, Elsevier Sciences, Ltd., 1996.

(26)     Hale, R., B. Kirwan, and F. Guldenmund, Capturing the River: Multilevel Modeling of Safety Management, Chapter 11, *Nuclear Safety: A Human Factors Perspective*, Misumi, Wilpert & Miller, Editors, Taylor & Francis, Ltd., 1999.

(27)     Marca, David A., and Clement L. McGowan, *SADT — Structured Analysis and Design Technique*, McGraw Hill, New York, 1986.

(28)     Osborn, R.N., J. Olson,, P.E. Sommers, S.D. McLaughlin, M.S. Jackson, W.G. Scott,, and P.E. Connor, "Organizational Analysis and Safety for Utilities with Nuclear Power Plants — Vol. 1, An Organizational Overview," NUREG/CR-3215, Pacific Northwest Laboratory, Prepared for U. S. Nuclear Regulatory Commission, August 1983.

(29)     Olson, J., S.D. McLaughlin, R.N. Osborn, and D.H. Jackson, "An Initial Empirical Analysis of Nuclear Power Plant Organization and Its Effect on Safety Performance," NUREG/CR-3737. Pacific Northwest Laboratory, Prepared for U. S. Nuclear Regulatory Commission, November 1984.

(30)     Marcus, A.A., M.L. Nichols, P. Bromiley, J. Olson, R.N. Osborn, W. Scott, P. Pelto, and J. Thurber, "Organization and Safety in Nuclear Power Plants," NUREG/CR-5437, Strategic Management Research Center, University of Minnesota, Prepared for U.S. Nuclear Regulatory Commission, May 1990.

(31)     Nichols, M.L., A.A. Marcus, J. Olson, R.N. Osborn, W. Scott, P. Pelto, J. Thurber, and G. McAvoy, "Organizational Factors Influencing Improvements in Nuclear Power Plants" (Draft), NUREG/CR-5705, Strategic Management Research Center, University of Minnesota, Prepared for U.S. Nuclear Regulatory Commission, October 9, 1992.

(32)     Haber, S.B., J.N. O'Brien, D.S. Metlay, and D.A. Crouch, "Influence of Organizational Factors on Performance Reliability," NUREG/CR-5538, Volume 1, Overview and Detailed Methodological Development, Brookhaven National Laboratory, Prepared for U.S. Nuclear Regulatory Commission, December 1991.

(33)     Mintzberg, Henry, *The Structuring of Organizations*, Prentice Hall, 1979.

(34)     Haber, Sonja B., Deborah A. Shurberg, Michael T. Barriere, and Robert E. Hall, "The Nuclear Organization and Management Analysis Concept Methodology: Four Years Later," 1992 IEEE Fifth Conference on Human Factors and Power Plants, Monterey, California, June 7–11, 1992.

(35)     Haber, Sonja B., Deborah A. Shurberg, and Michael T. Barriere, "Organizational Factors and Performance Reliability," Proceedings of the Probabilistic Safety Assessment International Topical Meeting, Clearwater Beach, Florida, January 26–29, 1993.

(36)     Jacobs, Rick, John Mathieu, Frank Landy, Tony Baratta, Gordon Robinson, David Hofmann, and Kathleen Ringenbach, "Organizational Processes and Nuclear Power Plant Safety," 1992 IEEE Fifth Conference on Human Factors and Power Plants, Monterey, California, June 7–11. 1992.

(37)     Jacobs, Rick and Sonja Haber, "Organizational Processes and Nuclear Power Plant Safety," *Reliability Engineering and System Safety*, 45, 75-83, Elsevier Science Limited, 1994.

(38)     Davoudian, Keyvan, Jya-Syin Wu, and George Apostolakis, "Incorporating Organizational Factors into Risk Assessment through the Analysis of Work Processes," *Reliability Engineering and System Safety*, 45, 85-105, 1994.

(39)     Davoudian, Keyvan, Jya-Syin Wu, and George Apostolakis, "The Work Process Analysis Model (WPAM)," *Reliability Engineering and System Safety*, 45, 107-125, 1994.

(40)     Apostolakis, G.E., "Organizational Factors and Nuclear Power Plant Safety," Second International Conference on Human Factors Research in Nuclear Power Operations, Berlin, 1996.

(41)     Marcinkowski, K., G. Apostolakis, and R. Weil, "A Computer Aided Technique for Identifying Latent Conditions (CATILaC)," *Cognition, Technology and Work*, 3, 111-126, Springer-Verlag London Limited, 2001.

(42)     Weil, Rick, and George Apostolakis, "On the Inclusion of Organizational Factors in Incident Investigation," PSA'99, Washington, D.C., 1999.

(43)     Weil, Rick, and George Apostolakis, "Identification of Important Organizational Factors Using Operating Experience," Presented at the 3rd International Conference on Human Factor Research in Nuclear Power Operations, Mihama, Japan, September 8–10, 1999.

(44)     Sewell, R.T., M. Khatib-Rahbar, and H. Erikson, "Implementation of a Risk-Based Performance Monitoring System for Nuclear Power Plants: Phase II - Type-D Indicators," ERI/SKi 99-401, February 1999.

(45)     Olson, J., R.N. Osborn, D.H. Jackson, and R. Shikiar, "Objective Indicators of Organizational Performance at Nuclear Power Plants," NUREG/CR-4378, January 1986.

(46)     Travers, William D., Memorandum to the Commissioners, "Status of Risk-Based Performance Indicator Development and Related Initiatives," SECY-00-0146, U.S. Nuclear Regulatory Commission, June 28, 2000.

(47)     Barriere, M.T., W.J. Luckas, Jr., D.A. Stock, and S.B. Haber, "Incorporating Organizational Factors into Human Error Probability Estimation and Probabilistic Risk Assessment" (Draft), Brookhaven National Laboratory, January 25, 1994.

(48)     Rogers, Kenneth C., "Progress Toward Risk-Informed Regulation," Speech given at the Twenty-Fourth Water Reactor Safety Information Meeting, Bethesda, MD, October 21, 1996.

(49)     ACSNI Study Group on Human Factors, "Third Report: Organizing for Safety," Advisory Committee on the Safety of Nuclear Installations, Health and Safety Executive, United Kingdom, 1993.

(50)     Berman, Jonathan, Philip Brabazon, Linda Bellamy, and Jo Huddleston, "The Regulator as a Determinant of the Safety Culture," prepared for the Health and Safety Executive, Nuclear Safety Research Management Unit, Four Elements Limited, London, September 1, 1994.

(51)     Marcus, Alfred A., "Implementing Externally Induced Innovations: A Comparison of Rule-Bound and Autonomous Approaches," *Academy of Management Journal*, Vol. 31, No. 2, 1988.

(52)     International Atomic Energy Agency, "Developing Safety Culture in Nuclear Activities: Practical Suggestions to Assist Progress," Safety Reports Series No. 11, Vienna, 1998.

(53)     International Nuclear Safety Advisory Group, "Basic Safety Principles for Nuclear Power Plants 75-INSAG-3 Rev. 1," INSAG-12, International Atomic Energy Agency, Vienna, 1999.

(54)     International Nuclear Safety Advisory Group, "Management of Operational Safety in Nuclear Power Plants," INSAG-13, International Atomic Energy Agency, Vienna, 1999.

(55)     Nuclear Energy Agency, "The Role of the Nuclear Regulator in Promoting and Evaluating Safety Culture," Organization for Economic Cooperation and Development, June 1999.

(56)     Nuclear Energy Agency, "Regulatory Response Strategies for Safety Culture Problems," Organization for Economic Cooperation and Development, 2000.

(57)     Nuclear Energy Agency, "Identification and Assessment of Organizational Factors Related to the Safety of NPPs," Organization for Economic Cooperation and Development, NEA/CSNI/R(99)21, September 1999.

(58)     Paradies, M., L. Unger, P. Haas, M. Terranova, "Development of the NRC's Human Performance Investigation Process (HPIP)," NUREG/CR-5455, U.S. Nuclear Regulatory Commission, 1993.

(59)     U.S. Nuclear Regulatory Commission, "Policy Statement On the Conduct of Nuclear Power Plant Operations," 54 FR 3424, January 24, 1989.

(60)     U.S. Nuclear Regulatory Commission, "Freedom of Employees in the Nuclear Industry to Raise Safety Concerns Without Fear of Retaliation," Policy Statement, 61 FR 24336, May 14, 1996.

(61)     U.S. Nuclear Regulatory Commission, Inspection Procedure 71841, "Human Performance," NRC Inspection Manual, December 12, 2000.

(62)     U.S. Nuclear Regulatory Commission, Inspection Report #72-1004/97-209, VECTRA Technologies, October 27 through November 6, 1997.

(63)     Travers, William D., Memorandum to the Commissioners, "Recommendations for Reactor Oversight Process Improvements," SECY-99-007, U.S. Nuclear Regulatory Commission, January 8, 1999.

(64)     Nishijima, Yoshimasa, "Human Performance Indicators," Chapter 14, *Nuclear Safety: A Human Factors Perspective*, Misumi, Wilpert, & Miller, Editors, Taylor & Francis, Ltd., 1999.

(65)     Pidgeon, Nick, "Safety Culture: Key Theoretical Issues," *Work and Stress*, Volume 12, Number 3, p. 202, Taylor & Francis, Ltd., July–September 1998.

(66)     Advisory Committee on Reactor Safeguards, Letter to Richard A. Meserve, Chairman, U.S. Nuclear Regulatory Commission, "SECY-00-0053, NRC Program on Human Performance in Nuclear Power Plant Safety," May 23, 2000.

## 10.2 References Not Cited

(1)    Barrier, M., W. Luckas, D. Whitehead, A. Ramey-Smith, "An Analysis of Operational Experience During Low Power and Shutdown and a Plan for Addressing Human Reliability Issues," NUREG/CR-6093, BNL-NUREG-52388, SAND93-1804, U.S. Nuclear Regulatory Commission, June 1994.

(2)    Brookhaven National Laboratory, "Organizational Factors Research to Support Accident Management" (Draft, transmitted by letter from Sonja B. Haber), NUREG/CR-6219, U.S. Nuclear Regulatory Commission, August 1993.

(3)    Callan, L. Joseph, Memorandum from the Executive Director for Operations, NRC, to the Commissioners, "Proposed Options for Assessing the Performance and Competency of Licensee Management," SECY-98-059, March 26, 1998.

(4)    Cummings, James J., Memorandum to the Commissioners, "Audit of NRC's Implementation of the TMI Action Plan," U.S. Nuclear Regulatory Commission, June 4, 1981.

(5)    Dircks, William J., Memorandum to the Commissioners, "Human Factors Program Plan," SECY-82-462, U.S. Nuclear Regulatory Commission, November 19, 1982.

(6)    Dougherty, Ed, "Human Errors of Commission Revisited: An Evaluation of the ATHEANA Approach," *Reliability Engineering and System Safety*, 60, p. 71–82, 1998.

(7)    Gertman, David I., Bruce P. Hallbert, Donald, L. Schurman, and Harold S. Blackman, "Management and Organizational Factors Technical Report: Socio-Organizational Contribution to Risk Assessment and the Technical Evaluation of Systems (SOCRATES)," NUREG/CR-6612 (Draft), U.S. Nuclear Regulatory Commission, Washington, D.C., March 1998.

(8)    Hoyle, John C., Memorandum from the Secretary of the Commission, NRC, to L. Joseph Callan, Executive Director for Operations, NRC, "Staff Requirements: SECY-98-059 — Proposed Options for Assessing the Performance and Competency of Licensee Management," June 29, 1998.

(9)    Kaufman, F., N. Todreas, and H. Arnold, "Report of the Independent Review Panel to Atomic Energy Control Board and Ontario Hydro," Volume 1, March 9, 1999.

(10)   Merrified, Jeffrey S., Remarks at the Regulatory Information Conference, Washington, D.C., March 29, 2000.

(11)   Olson, J., A.D. Chockie, C.L. Geisendorfer, "Development of Programmatic Performance Indicators," NUREG/CR-5241, PNL-6680, U.S. Nuclear Regulatory Commission, October 1988.

(12)   Perrow, Charles, *Normal Accidents: Living with High Risk Technologies*, Basic Books, Inc., New York, 1984.

(13)   Peters, Thomas, and Robert H. Waterman, Jr., *In Search of Excellence*, Harper & Rowe, 1982.

(14)   Reason, J., "Are We Casting the Net Too Widely in Our Search for the Factors Contributing to Errors and Accidents?", *Nuclear Safety: A Human Factors Perspective*, Misumi, Wilpert and Miller, Eds., Taylor & Francis, Ltd., London, 1999.

(15)   Reason, James, "Achieving a Safe Culture: Theory and Practice," *Work and Stress*, Volume 12, Number 3, p. 293, Taylor & Francis, Ltd., July–September 1998.

(16)   Ryan, Thomas G., "Organizational Factors Research Lessons Learned and Findings," Office of Nuclear Regulatory Research, U.S. Nuclear Regulatory Commission, 1991.

(17)   Shurberg, Deborah, and Sonja B. Haber, "Techniques to Assess Organization Factors: Progress to Date," BNL Technical Report A-3956-1-7/94, Brookhaven National Laboratory, July 1994.

(18)   Shurberg, Deborah, Sonja B. Haber, and David Hofmann, "Results of a Pilot Application for Rating Organizational Performance Factors Based on Analysis of Existing Documentation (Draft)," Brookhaven National Laboratory, prepared for U.S. Nuclear Regulatory Commission, July 1995.

(19)   Strater, Oliver, and Bubb Heiner, "Assessment of Human Reliability Based on Evaluation of Plant Experiences: Requirements and Implementation," *Reliability Engineering and System Safety*, 63, 199–219, 1999.

(20)   Taylor, James M., Memorandum from the Executive Director for Operations, NRC, to the Commissioners, "Review of Organizational Factors Research," SECY-93-020, February 1, 1993.

(21)   Thurber, J.A., J. Olson, R.N. Osborn, P. Sommers, and R.D. Widrig, "Guidelines and Workbook for Assessment of Organization and Administration of Utilities Seeking Operating License for a Nuclear Power Plant," NUREG/CR-4125, September 1986.

| NRC FORM 335<br>(2-89)<br>NRCM 1102,<br>3201, 3202 | **U.S. NUCLEAR REGULATORY COMMISSION**<br><br>**BIBLIOGRAPHIC DATA SHEET**<br>*(See instructions on the reverse)* | 1. REPORT NUMBER<br>(Assigned by NRC, Add Vol., Supp., Rev.,<br>and Addendum Numbers, if any.)<br><br>NUREG-1756 |
|---|---|---|

**2. TITLE AND SUBTITLE**

SAFETY CULTURE: A SURVEY OF THE STATE-OF-THE-ART

Prepared for the Advisory Committee on Reactor Safeguards

| 3. | DATE REPORT PUBLISHED |
|---|---|
| MONTH | YEAR |
| January | 2002 |

4. FIN OR GRANT NUMBER

**5. AUTHOR(S)**

J. N. Sorensen
Senior Fellow

6. TYPE OF REPORT

Technical

7. PERIOD COVERED *(Inclusive Dates)*

**8. PERFORMING ORGANIZATION - NAME AND ADDRESS** *(If NRC, provide Division, Office or Region, U.S. Nuclear Regulatory Commission, and mailing address; if contractor, provide name and mailing address.)*

Advisory Committee on Reactor Safeguards
U. S. Nuclear Regulatory Commission
Washington, DC 20555-0001

**9. SPONSORING ORGANIZATION - NAME AND ADDRESS** *(If NRC, type "Same as above"; if contractor, provide NRC Division, Office or Region, U.S. Nuclear Regulatory Commission, and mailing address.)*

Same as above

**10. SUPPLEMENTARY NOTES**

**11. ABSTRACT** *(200 words or less)*

This report was prepared for the Advisory Committee on Reactor Safeguards to provide background information on the evolution of the term "safety culture" and the perceived relationship between safety culture and the safety of operations in nuclear power generation and other hazardous technologies. There is a widespread belief that safety culture is an important contributor to the safety of operations. Empirical evidence that safety culture and other management and organizational factors influence operational safety is more readily available for the chemical process industry than for nuclear power plant operations. The commonly accepted attributes of safety culture include good organizational communication, good organizational learning, and senior management commitment to safety. Safety culture may be particularly important in reducing latent errors in complex, well-defended systems. The role of regulatory bodies in fostering strong safety cultures remains unclear, and additional work is required to define the essential attributes of safety culture and to identify reliable performance indicators.

**12. KEY WORDS/DESCRIPTORS** *(List words or phrases that will assist researchers in locating the report.)*

Human factors
Human performance
Managment and organizational factors
Performance indicators
Safety culture
Safety management

| 13. AVAILABILITY STATEMENT |
|---|
| unlimited |
| 14. SECURITY CLASSIFICATION |
| *(This Page)* |
| unclassified |
| *(This Report)* |
| unclassified |
| 15. NUMBER OF PAGES |
| 16. PRICE |

This form was electronically produced by Elite Federal Forms, Inc.

NUREG-1756

SAFETY CULTURE: A SURVEY OF THE STATE-OF-THE-ART

JANUARY 2002